WHAT IS GRAVITY?

WHAT IS GRAVITY?

Eugene E. Bunt

VANTAGE PRESS
New York

FIRST EDITION

Published by Vantage Press, Inc.
516 West 34th Street, New York, New York 10001

Manufactured in the United States of America
ISBN: 0-533-12377-1

Library of Congress Catalog Card No.: 97-90374

0 9 8 7 6 5 4 3 2 1

This book is dedicated to my wife's grandfather, James Grandville Pegg, whose enthusiasm to have an author in the family kept me going. Also, Dr. James McKenna, for all his encouragement and support. A special thanks to my family: my wife, Patti, and two children, Eric and Nichole, who allowed me the time to write, and to the rest of the family, thank you, for listening and supporting me.

Many thanks to all,
Eugene E. Bunt

Contents

Preface		ix
One.	History	1
Two.	Background	11
Three.	Light	19
Four.	Gravity	35
Five.	Laser	59
Six.	The Bomb	67
Seven.	Saturn	77
Eight.	Tesla	79
Nine.	Electrons	86
Ten.	The Beginning of the Universe	107
Eleven.	Conclusion	115

Preface

This is a nonconventional theory of light and gravity. Light is a two-dimensional phenomenon, possessing only length and height, exclusive of a time element. Gravity is the result of a distortion of these dimensions relative to each other.

I have based my ideas on experiments that have already been performed. I have also tried to imaginatively associate them in such a way as to lead away from prevailing theorems. I found as I wrote them that some of the ideas strain even my credulity, but I wanted to test the boundaries of this trend of thought and still try to be realistic. I find the basic premise more plausible than current theories. If any of these proposed ideas prove to be true, the implications are staggering. This is an exercise in imagination, and I hope to stimulate some new ideas.

There are problems that arise when you attempt to introduce a new basic theory into physics when all previous theories have been interrelated. The theories that were conceived after the idea you wish to challenge were interpreted to conform to that very theory you are replacing. In order to change one aspect, you also have to redefine the relation of a multitude of well-entrenched theories. There are major problems with current theories that need resolving.

I enjoyed reading an anecdote about E. O. Lawrence, who, after trying some intricate method of building his cyclotron, and failing, remarked to his assistant, "Good! That eliminates that."

One
History

Man has been struggling to understand the basic functions of the universe through all recorded history.

Over two thousand years ago the Greeks were contemplating the discontinuity of motion. The Greek thinkers Zeno and Aristotle point out the difficulty of analyzing motion as a series of still frames. One hundred years before Aristotle, Zeno from Elea speculated on the true nature of motion. He stated his ideas as three paradoxes.

Zeno's first paradox:

> Motion cannot exist, because in order for a runner to reach the endpoint of his race, he must first reach the point halfway between the endpoint and the starting point. But don't you see the paradox in this? Before he reaches the halfway point, he must reach the point midway between the halfway point and the starting line. Well, there is a problem, because the statement I just made can be applied to any distance in this race. Before a runner can reach the quarter point, he must first reach the one-eighth point and before that, the one-sixteenth point, etc. . . .

That's it! There are an infinite number of these halfway points, and each of them marks off a finite distance the runner must reach before he is able to go on to the next point. Therefore, there is no first distance he can run to, because. . . .

Zeno's second paradox follows the logic of his first:

Achilles can never overtake the tortoise in a race, even though the slower tortoise is ahead of Achilles by a modest distance, the faster Achilles cannot reach him. Because to do so, Achilles must first reach the point just departed by the turtle before he can catch up to the turtle. As my first paradox demonstrated, there would be an infinite number of such points for Achilles to face. Certainly as long as there is any distance between them, the turtle will have left any point before Achilles arrives there.

Zeno's third paradox:

The arrow cannot fly. It cannot fly because an object that is behaving itself in such a uniform manner is either continually moving, or continually at rest. The arrow is certainly behaving itself in such a uniform manner. Now watch the arrow as it travels along its flight path. Clearly, at any instant, the arrow is occupying a given place. Therefore, if it is occupying a place, it must be at rest there. The arrow must be at rest at the instant we picture, and since the instant we have chosen is any instant, the arrow cannot be moving at any instant. Thus the arrow is always at rest and cannot fly.

Life's experiences do not coincide with our logical deductions. Faster runners overtake slower-moving runners and arrows fly. Faced with the obvious, Aristotle, who considered Zeno's ideas important, felt the answer was the wholeness of continuity and the mistrust of analysis. He presented Zeno's paradoxes and his own resolutions in *Physica.* These ideas had been neglected until Max Planck's formulae predicted the quantum characteristics of energy emitted by atoms. On an atomic scale, particles jump instantly from one spot to another without passing through any spot in between.

The next really bright spot in the history of science was caused by the genius of Galileo. Galileo was the forerunner of the modern physicist. Not only did he gaze into the solar system with his invention, the telescope, but he also devised methods of observation, description, and analysis that we take as the basis for all physics. What is viewed as Galileo's most important contribution is his observation pertaining to falling objects. Up until Galileo's time, it was assumed that heavy objects fell faster than light ones. It was Galileo who deduced that if you could overcome air resistance, all objects would accelerate and fall at the same rate. This is truly remarkable when you consider that an object's gravitational mass is equivalent to its acceleration mass. This also holds true in lesser gravitational conditions; Buzz Aldrin did Galileo's experiment on the Moon with the same results.

The Catholic Church censored scientific thinking during the time of Copernicus. Copernicus wrote that the Earth might not occupy the center of the universe but may be traveling through it. This statement, during his period of time, was heresy. His book *De Revolutionibus Orbium Caelestium* (The Revolution of the Heavenly Spheres), which was printed shortly before his death, was banned by the church and not seen again for three hundred years. Somehow, Giordano Bruno heard of Copernicus' theories. Bruno took it upon himself to spread these ideas freely, for which he was burned at the stake as a heretic in the year 1600.

Johannes Kepler, working with Tycho Brahe as his assistant, set out to confirm Nicolaus Copernicus' view that the Earth orbited the Sun. Their three laws for celestial motion were the foundation for Sir Isaac Newton's law of motion and law of universal gravitation. Newton's ideas were limited to all things sharing the same time and length scales. In his time, they had not developed instruments accurate enough to detect any minute differences.

Newton also invented calculus in order to more easily predict the path of the planets in his laws of motion and theory of gravity. Newton developed a particle theory of light, and he performed experiments such as splitting light into a spectrum in one prism, then recombining it back into white light in a second prism.

Another person who greatly influenced Newton and whose writings had a major effect on quantum mechanics was René Descartes, whose famous postulation "I think, therefore I am" is still almost universally known today.

In other realms of science, Luigi Galvani discovered electricity in the late 1770s. The basic concepts were developed in friendly arguments between Galvani and Alessandro Volta.

Hans Christian Orsted expanded further on these ideas and experimentally linked electricity to magnetism, two phenomena that were formerly considered unrelated. The astounding part of this relationship is that unlike gravitation fields, which pull on a direct line between the heavenly bodies, the bar magnet points at right angles to the flow of the electric current.

Christian Huygens, a contemporary of Newton, put forward a wave theory of light that contradicted Newton's particle model. Huygens' theory was later substantiated in the year 1802 with a light-diffraction experiment performed by Thomas Young.

I'm changing subjects here, but I'm trying to keep the events somewhat in chronological order. Around 1820, a botanist by the name of Robert Brown, while studying pollen, discovered that small particles of inorganic matter remain in motion while suspended in water, even in the absence of any current. The existence of a never-diminishing motion seemed to contradict all previous experience and wasn't explained until after the turn of the twentieth century with the elastic bouncing atom model for gases and liquids. This motion has since been tagged with the name Brownian motion.

The understanding of electricity was expanded by experiments done by Henry Augustus Rowland in the mid–eighteenth century. He found that a moving magnet generated electricity and that the speed of the magnet affected the voltage. Rowland's work—along with experiments done by Charles Augustine Coulomb, Joseph Henry, Michael Faraday, and James Clerk Maxwell, whose equations described the electromagnetic field—set the stage for present electrical theories.

Nikola Tesla's research was almost completely censored because of the great animosity existing between himself and Thomas Edison. Edison, who had great political clout through his friendship with Henry Ford, managed to discredit Tesla.

In spite of Edison, Tesla found backing for his experiments from Westinghouse. Edison believed that direct current was the most efficient form of electricity whereas Tesla proved that alternating current was more conducive to transmission through wires. If it had been left to Edison, we still wouldn't have buildings over ten stories high and we would have to build power stations every quarter mile.

The wealth of Tesla's discoveries are still being uncovered, primarily through a library set up in his honor in his native land, Serbia. He took X rays before Roentgen, but his lab was burned down along with all his experimental results. He also displayed a radio-controlled boat in New York harbor several years before Marconi invented the radio. His electrical paraphernalia was so astounding, they were used as backdrops in the first *Frankenstein* movie, which is still a thrill to watch.

I've given Tesla additional recognition because I feel that he has been maligned. His contributions—from neon lights and transformers to the use of microwave energy—have been overlooked and are still not credited to him.

In the 1880s and through the 1890s, at about the same time as the major contributions of Tesla, Albert Abraham Michelson and Edward Williams Morley tried to experimentally establish the speed of the Earth by using the constant speed of light as the means to determine what stationary should be.

At the turn of the century, a real flurry of theories and models were put forward. We still use many of them today for lack of anything better with which to replace them.

Jean Baptiste Perrin found substantial proof that cathode rays were negatively charged particles. The word *electron* was suggested by George Johnstone Stoney in 1894. Michael Faraday had attempted to influence light with magnetism by having light pass through a prism in a magnetic field. These attempts led him to the important discovery of the magnetic field rotating the plane of polarization of light, passing through the affected glass.

It wasn't until 1896 that Pieter Zeeman succeeded in twinning spectral lines using a diffraction grating immersed in a magnetic field. Pieter Zeeman used sodium light for its purity. (Sodium emits only yellow light and only has two spectral lines.) Thus magnetism was shown to have an influence on light.

The year 1896 also saw the discovery of radioactivity by Henry Becquerel. Madame Curie followed up on his discovery and coined the word *radioactivity.* Madame Curie and her husband established the fact that more than one element is radioactive. The Curies also discovered that radioactive elements had a *half-life.*

Ernest Rutherford used radioactive substances to experimentally establish the locations of the constituent parts of the atom by bombarding a leaf of gold foil with protons.

Around 1900 an obscure professor named Max Planck discovered while studying thermodynamics that radiation is emitted at specific levels of energy. This discovery corresponded to Zeno's still-frame vision of movement. I will elaborate more on this later.

While pondering the duplicity of the Michelson-Morley experiment, which concluded that there was no difference in the amount of time it took light to travel in directions perpendicular to each other, Hendrik Lorentz speculated on a possible solution to this problem, a solution that was later called the "Lorentz Transformation." Without some sort of modification, given Michelson-Morley's results, the Earth would have to be immobile at the center of a moving universe. The Lorentz Transformation theory stated that objects had to shrink in the direction of their motion. This theory almost compensates for light taking the same amount of time to travel in opposing directions, one of which was visibly determined to coincide with the motion of the Earth through the universe. As unlikely as this seems, something has to be altered in order for reality to match experiment, and even this adjustment was not enough to cover the inequities.

Lorentz's theory was quickly followed by Einstein's "Theory of Relativity," which seems to have been an independent conclusion based on Maxwell's equations. The equations would have to hold true for any coordinate system, whether it is moving or not. The theory of relativity states that, instead of one shared time for the entire universe, each body, by virtue of either its speed or proximity to mass, changes its time-rhythm relative to the location of the observer. The time-rhythm slows as the speed of the body increases, and time slows with respect to distance as you approach any mass.

Einstein's predictions of time differences have been proven. Recently more than one research group has measured the discrepancy of time between two atomic clocks with as little difference in elevation as that offered by a ten-story building. The difference is infinitesimal; therefore, you need the accuracy of an atomic clock to notice the change. (An example of an atomic clock would be quartz. Quartz oscillates at a given frequency when subjected to a small voltage. The frequency is so universally predictable that quartz crystals are at the heart of some of the finest timepieces in the world.)

Observable proof that conditions alter time scales, so that several comparable time scales exist, would be the rotation of galaxies. Take any bar- or spiral-type galaxy as an example; galaxies must rotate as if the star's positions were marked on this piece of paper—then rotate the paper. Galaxies have to rotate as a solid unit, or a galaxy would lose its shape within a single rotation.

Compare galactic rotation to our solar system's planetary movement with the different orbital periods for each planet. At greater distances from the Sun, the planets have to possess greater velocities in order to maintain their more remote orbits. So must the outer stars in the galaxies. The way out of this dilemma is to assign separate time zones, with regards to the distance from the center of mass, for each section of the galaxy being considered.

The orbiting electron model put forward by Neils Bohr is no longer accepted as reality, but is still the textbook example. The diameter of the Bohr orbit was derived from a multiple of Planck's formula. The electron can be at one level or the next, but can never exist, even for an instant, between levels.

Louis de Broglie found that the electron also possessed a wave. Erwin Schrodinger calculated the size and the frequency of that wave in relation to the electron's distance from the nucleus. An astonishing feature of the Schrodinger wave function is that all of the electrons in an atom cycle in a synchronized fashion.

The astounding experimental confirmation of these outlandish ideas caused the Cartesian philosophy to crest at the fifth Solvay Congress, held in 1927, which was privately sponsored by Belgian industrialist Ernest Solvay.[1] After dividing the scientists into two debating teams led by Einstein on the one side and Bohr, Heisenburg, and Schrodinger on the other, the Cartesian ideas prevailed, because the experimental results seemed so magical.

I think it is very self-centered to imagine that we control the universe by our observation of it. I believe Schrodinger's cat is alive or dead, regardless of what we think. For those who don't know, Schrodinger's cat was the hypothetical conclusion of the fifth Solvay Congress, on which all of modern physics is based. Schrodinger's cat is the syllogism that every time a choice presents itself, two or more universes pop into existence, one for every possible result of the event. The analogy goes that the live cat is placed into a box with a poison capsule. The capsule may or may not dissolve, killing or not killing the cat. They say both possibilities actually occur, causing two of you, one seeing a dead cat and the other a live cat, in two separate universes. In order to understand quantum physics, those who teach it have to prepare you mentally for some pretty unusual thought. If you jump right into those ideas without preparation you would automatically reject them as absurd. Here is a general synopsis of their results:

The Heisenburg-Schrodinger debating team's premise is based on the experimental results that you can either determine an electron's speed or location, but not both, and that depends on which experiment you perform first. They assumed that the experimenter's psyche played a role in obtaining the results by choosing which experiment was to be performed first. The dual nature of light also caused some confusion to the physicists of this period, prior to the debate. Part of the mystery is caused by the fact that light has to act as a particle in order to transit empty space, then as a wave (as has been demonstrated) when interacting with matter. Schrodinger's equations added to the mystery. Schrodinger's equations described the length of an electron's wave; the description only allowed one wave function for all the electrons of an atom. How can electrons, at varying shell (orbit) distances from the nucleus, all move varying lengths at the same instant? Because of all the incongruities in the laws of physics, the fifth Solvay Congress concluded that the observer controls what he is going to see happen.

These experimental results have since been updated by tracing particles in liquid hydrogen; now, both the speed and location of a particle can be determined at the same time. Before the liquid hydrogen experiment, using primitive equipment, experiments interfered with the results. These recent developments haven't brought any changes to current theories yet.

In order to change your opinion on present theories, I have to provide a viable alternative to the hypothesis that an observer interferes with the object observed. I'm not opposed to all the conclusions reached at the Solvay Congress. I do think that it's unreasonable for each individual to control a personal universe by virtue of occupying it, and I don't believe that the universe divides every time a choice presents itself.

Note

1. Fred Alan Wolf, *Taking the Quantum Leap* (New York: Harper & Row, 1981).

Two

Background

When light passes through a small aperture, it leaves a ring pattern on a wall (or projection screen). When light passes through two small holes side by side, it leaves a vertical line pattern, a typical wave-interference pattern. This experiment was instrumental in replacing Newton's corpuscular light theory with Huygens' wave theory because light is obviously a wave, bending around objects, creating wave-interference patterns (see Figures 1 and 2).

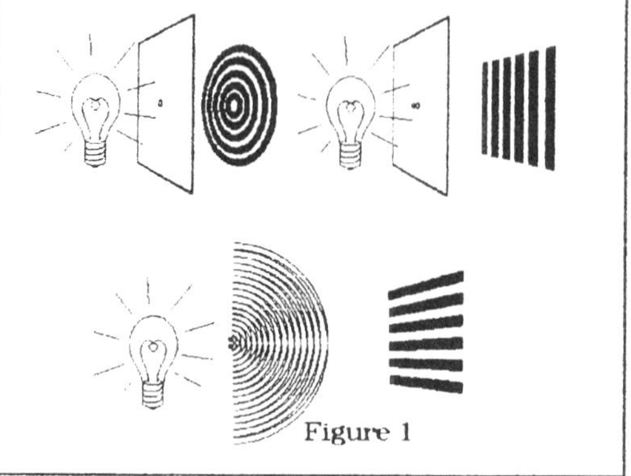
Figure 1

Figure 2

As illustrated in Figure 1 light cancels light in wave troughs and reinforces itself on wave crests. As depicted in Figure 2, the wave pattern exhibited in the two-hole diffraction pattern is enhanced when the light source uses a hyperbolically curved surface to focus the light, such as light from an auto headlamp. All rays of light from this configuration exit parallel to each other.

What could an ether consist of that will transmit a light-wave, be invisible, and not affect the motion of celestial bodies? The ether was an invention to explain the transmission of light through empty space. It is illogical to imagine light traversing empty space as a wave without something to conduct it. The problem is to design this ether so that light will travel through it like a wave through Jell-O, and leave matter unaffected. Light traveling through space has to have some characteristics. Light reacts with matter by heating it. It seems unrealistic that matter and this ether cannot be found to react with each other. Other problems that were encountered while formulating a theory for light were that you would expect light to be a longitudinal wave that moves similar to a sound-wave through air; the air is compressed into a thick wave front followed by a rarified segment of air. Contrary to what we expected to find, experiments showed that light is a transverse wave that moves in an up-and-down pattern. If it did not, we would be unable to explain how light is polarized and analyzed.

A transverse wave moves up and down and will pass through an analyzer only if it is properly aligned. To visualize this, imagine that you have two plates representing the analyzers. One has a vertical slit, while the other has a horizontal slit. Now imagine a rope vibrating on one plane—say it's moving up and down. This representation of a wave can pass through the vertical slit, but will not pass through the horizontal one (see Figure 3).[1]

Take light that has passed through a tourmaline crystal, for example. If you take the tourmaline crystal and split it in half and hold both halves in line as they came apart, light will shine through both halves. When you start to rotate one of the halves, the light will fade to nothing when you come to ninety degrees out of alignment. If you keep rotating the crystals until they reach 180 degrees, the light will gradually return to full brightness. Some materials will require lesser rotation angles—say, sixty degrees—and some more.

When light shines on a metal plate, the light dislodges electrons. The intensity of the light doesn't control the electrons' energy while the frequency of the light does. This lends credence to Newton's theory of light as particles. Following the logic of the wave theory, you would assume, by increasing the frequency of

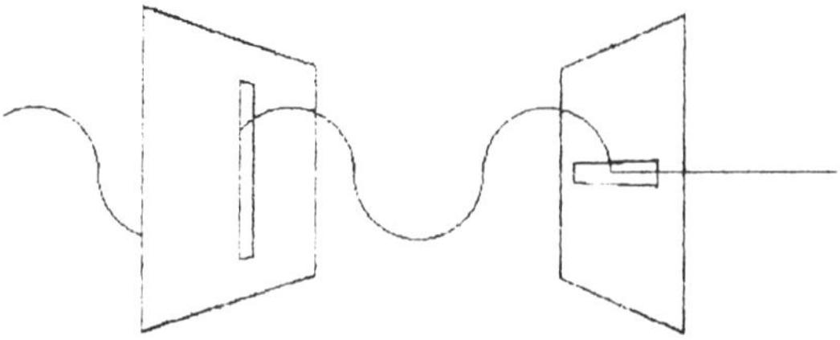

Figure 3

the light-wave striking a metal plate, a greater quantity of electrons would be carried off. Similarly, bringing the light source closer—or, in other words, increasing the intensity—should increase the potential of the electrons, which is not the case. Increasing the frequency somehow affects the impact of the photon, causing it to impart greater energy to the electrons, even though it travels with the same velocity as any other photon. The only way to make sense of this development is to attribute greater mass to the higher-frequency photon, which means it has to have a particle nature.

The two seemingly incompatible theories have been used in conjunction with each other in order to explain the dual nature of light.

James Clark Maxwell performed experiments establishing the speed of a magnetic wave to be equal to that of light. Maxwell also established that a radio signal's mode of travel is not a straight line, but one that keeps spiraling at right angles to itself. With the speed of a magnetic wave being equal to that of light, and light having been established to be a wave, the two phenomena that were formerly considered to be independent forms of energy are now thought to be different manifestations of the same phenomenon. The difference between light and a magnetic wave is that light has a much shorter wavelength.

There is nothing in everyday life to prepare the reader for the concept of relativity. The best I can do to bring an initiate to understanding it is state the reasons that led to the need for its conception.

Almost half of the stars in the sky are binary star systems. A binary star system is comprised of two stars that orbit around their common center of mass. In some of these systems, one of the two stars is massive enough to dominate the system, causing the other to maintain a planetary-type orbit. These stars have given astronomers the opportunity to study the effects of motion on light. Astronomers found that light always travels at the same speed, regardless of the speed or direction of travel of the emitting body.

Back in 1887, A. A. Michelson and E. W. Morley performed an experiment using the constant speed of light as a gauge to ascertain the absolute speed and direction of the Earth through the ether. Our planet is revolving around the Sun and our solar system is located approximately two/thirds of the distance outward on one of the rotating spiral arms in the Milky Way galaxy. The experiment was set up to measure and compare the speed of light in different directions. It was calculated that a slightly different length of time would be required for light to travel between two mirrors perpendicular to the relative motion of ether as opposed to the time it should take light to travel parallel to that motion. The startling results were that the speed of light was the same regardless of in which direction it was measured at any time of day or season of the year.

Given the fact that the speed of light emitted from a star is the same when it is traveling toward you as when it is retreating from you, and given the complete failure to detect a difference in the speed of light here on Earth, either the Earth is standing still at the center of the universe, or something else is happening.

Hendrik Lorentz's attempt to explain the results of the Michelson-Morley experiment with the Lorentz Transformation theory (stating that objects shrink in the direction of their motion), was ingenious, but it still didn't account for all of the experimental discrepancies. Einstein suggested in 1905 that you could do away entirely with the ether theory if you were willing to give up the notion of absolute time.

Let's dispense with the notion of simultaneous events in two coordinate systems that are moving relative to each other. Einstein's train example is the best analogy I know of:

We have two observers, one of whom is sitting on a railroad embankment, the other riding on a railcar traveling past the first observer. The speed of light is finite, so that in order to have the flash from two lightning strikes to converge at the same location adjacent to each other in the two coordinate systems that are moving relative to each other, the starting points can't be equidistant from both observers in each respective coordinate system. The two lightning strikes that the stationary man perceives as simultaneous appear as strike B before strike A to the man on the railcar, due to the car's motion. In order for the light to converge when the two observers are opposite each other, the light cannot start from equidistant points in both coordinate systems (see Figure 4).

The maximum speed that any influence can travel, be it light, gravity, or electromagnetic waves, is the speed of light. So any consideration on distant objects other than what is currently observed by you is irrelevant to you in your coordinate system.

Along with the theory of relativity, Einstein predicted that even light from the background stars would be bent inward toward the Sun, by a few degrees, from the Sun's gravitational pull. Scientists tried to verify that by photographing the Sun during a solar eclipse and comparing the positions of the background stars with existing photos of the same constellations. Instead of being pulled inward toward the Sun, as predicted, they

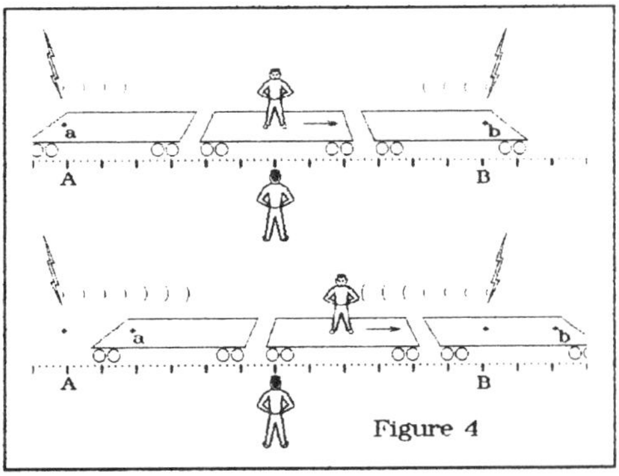

Figure 4

were bent away, so the explanation had to be modified. The new version stated that the light passing through the space immediately outside the space containing the light directed toward us was pulled in to give the appearance of being bent away from the Sun. During a solar eclipse, the background stars appear to move away from the Sun in relation to their normal position, as their light passes close to the Sun's gravitational field (see Figure 5).

This also gave a reason for the uneven orbit of Mercury, which astronomers had known about for some time. According to Newtonian physics, the orbit should be the same on every pass, but due to the difference in its time structure it appears to travel with less velocity than it should obtain at perihelion, causing the apogee of its orbit to move into a different position on every pass in a rosette pattern (see Figure 6).

One day on Mercury lasts fifty-nine days, two-thirds of its eighty-eight-day year. At perihelion, Mercury's rotation slows so that it revolves on its axis more slowly than it orbits the Sun, so that a person standing on Mercury would actually observe the Sun stopping in its path and traveling east before returning to its normal westward procession.[2] If this slowing of the planet's

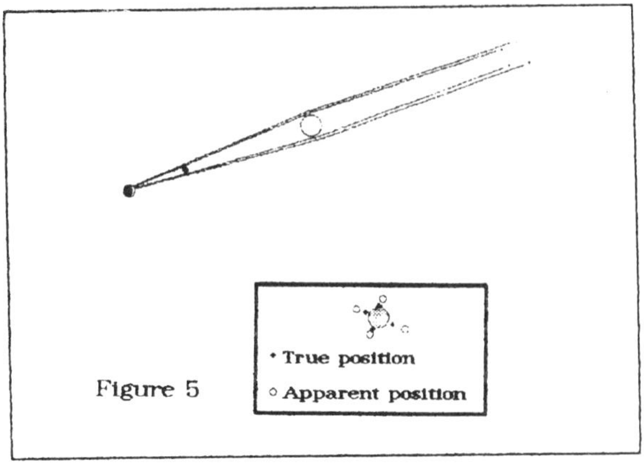

Figure 5

rotation were due to the Sun's gravitational pull on an equatorial bulge, the bulge would have to be migratory in order for this to happen repeatedly. Mercury's daily rotation period is not equal to its year.

Notes

1. Elizabeth A. Wood, *Crystals and Light* (Princeton, NJ: D. Van Nostrand, 1964).
2. Eric Flaum, *The Planets: A Journey into Space* (New York: Michael Friedman, 1988).

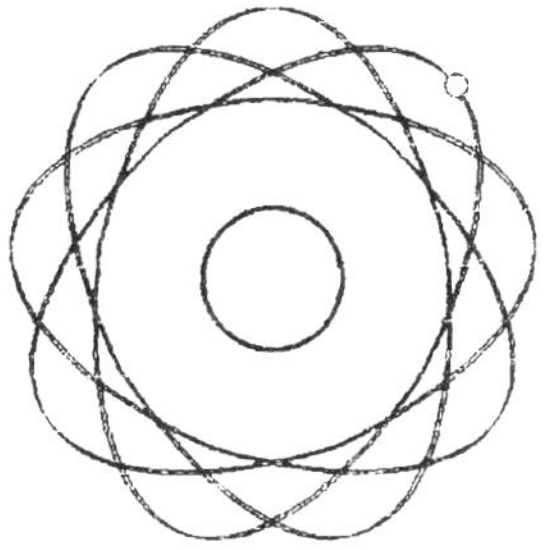

Figure 6

Three
Light

Sir Isaac Newton put forward a corpuscular theory for light. He theorized that light was made of tiny particles. Newton knew that when you mix small grains of different colored sand they appear to be a different color than either of the original colors. These particles in theory had a different mass for different colors, which affected the angle of penetration for each color going into a new medium. Newton also demonstrated that light, after having been separated into its spectrum in one prism, can be recombined back into white light in a second prism.

Huygens put forward a wave theory of light, which confronted the division of light into its spectrum as if light acted like a row of marching soldiers locked into a set distance from each other. As the line made a turn into a new medium, it took longer for the lengthier rows to attain the new angle. Light rays of different colors, running nearly parallel to each other, interfere with each others' waves. This interference causes a light ray with a wave characteristic of a third color to materialize. Light rays that intersect at wide angles do not interfere with each other. At the time they were proposed, both Newton's and Huygens' theories fit the problem equally well for separating colors through a prism. Newton's particle theory of light was finally dispatched by Thomas Young's wave interference experiment.

If light is a wave, as proven by the Thomas Young experiment, then the order of light exiting from a prism makes no sense. As the light entering the new medium attains a different velocity, the light pivots on its initial contact point. The degree of the new angle should depend on the height of the wave. The color blue being of shorter wavelength, you would assume it would also be shorter in height, making the smaller turn. Red by the same token should be taller and make the wider turn. With the color red making the greater turn, the spectrum would be inverted from any illustration that I ever saw. I followed this up with a few observations and concluded that the logical solution using the wave theory is an illusion (see Figure 7).

Popular theory states that the mass of a photon is so minute that it cannot be directly measured. Conclusions drawn from experiments using a photocell indicate by means of deductive reasoning that photons of shorter wavelength have greater mass. When Einstein postulated that a photon had a dual nature, and acted as a particle at times, it was due in part to the way light acts in a photocell. Einstein's conclusions on the photon were that it (a) acts as a wave in the presence of the matter from which it radiates, (b) becomes a particle in order to transverse the empty reaches of space, and (c) returns to its wave form when it comes in contact with matter again.

If light were a wave, you would assume that when you shine

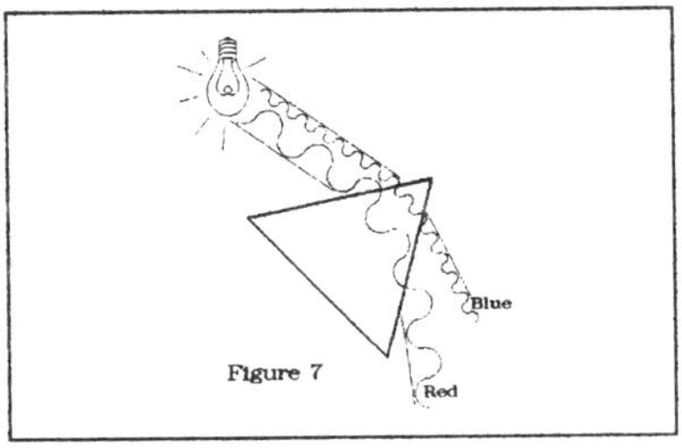

Figure 7

light on a photocell from a monochromatic light source, you would derive more energy—or *voltage*—from the electrons by merely increasing the intensity of the light striking the surface. Using the wave theory, logically, the frequency of the wave should control the volume or *amperage*.

If photons were composed of massless particles that were all traveling with the same velocity, they should also be identical in inertial energy. Higher frequency ultraviolet light gives greater electrical potential differences than the red end of the spectrum. Henry Augustus Rowland found that by increasing the speed of a magnet in the presence of a wire, there is an increase in voltage within the wire—speed is a factor in determining voltage. On striking the electron, the higher frequency photon causes the electron to bounce off at a greater velocity, thus causing an increase in voltage. Increasing the intensity of the light allows a greater volume of electrons—or amperage—with no increase in electron potential. Each photon acts as if it were a bullet of predetermined energy relative to its frequency. What else can you deduce other than that the higher frequency light-particles—photons—possess greater mass?

When light falls on the photocell, it travels through the film and the barrier, striking the light-sensitive selenium. The selenium releases electrons that travel through the barrier (which will only allow electrons to travel in one direction) to collect on the thin transparent metallic film. The copper ring that is pressed to the film is hooked to a wire that carries the negative charge through the circuit. The typical photocell is constructed as shown in Figure 8.

Newton's corpuscular theory of light also has problems explaining the angle of deflection for the different colors. Blue photons, being more massive, should experience less deviation from their initial direction, making a smaller turn prism. This would leave blue at the top of the spectrum, exiting from where the prism is thinnest. This expected order of colors is also

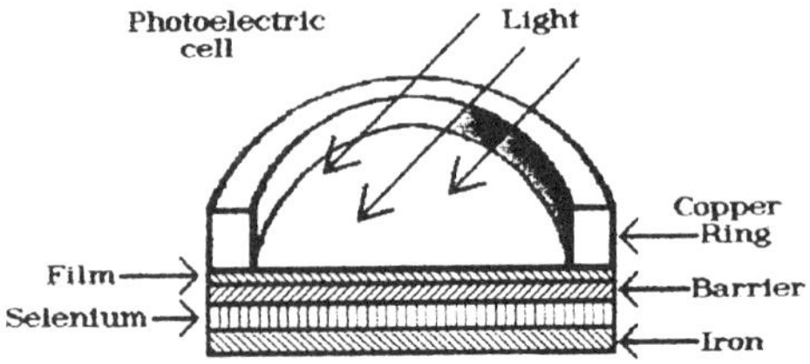

Figure 8

reversed from the true order of colors of the spectrum exiting from a prism.

Light energy is assimilated by an atom through its electrons. Electrons are located at certain distances from the nucleus. These distances have been designated as shells. When a photon collides with an atom, it is absorbed by an electron, and the electron raises one or more shell levels from the nucleus. The new shell that it occupies represents the combined energy of the photon added to the electron energy level of the departed shell. Conversely, the atom disperses a photon when the electron drops from a higher shell to a lower one. The photon produced will reflect the amount of energy the electron loses.

The electrons have been shown to exist in shells. These shells are elliptically shaped and are assumed to be located at fixed intervals that can be determined using Bohr's formula. When an atom absorbs energy, the electron jumps from an inner to an outer shell. The energy that an atom can acquire is restricted to the availability of an electron capable of using that amount of energy to jump to an outer shell. In the same manner, the electron emits a photon when the electron jumps down a shell level. The electron can accept energy and jump one, two, or more shells outward, then give up the acquired energy in steps or all at once. It's interesting to look at them on a larger scale and compare them to other phenomena (see tables 1, 2 and 3). The electron can also absorb energy in small steps, then send it off in one large jump.

The shells and the number of electrons required to fill them can be calculated. The shells are full when they contain (n^2) times two electrons, where n equals the shell number starting from the inner shell (2, 8, 18, 32). In addition, the shells are further divided into subshells. These subshells can contain 2, 6, 10, and 14 electrons respectively. Beyond the fourth shell, the numbers of electrons has not been experimentally established.

Table 1

$Number^2$	Number × 2	Electrons per shell	Products added	Electrons per subshell	Subshell divided by 2
1	2	2	+ 6 = 8	2	/ 2 = 1
4	8	8	+ 10 = 18	6	\ 2 = 3
9	18	18	+ 14 = 32	10	/ 2 = 5
16	32	32	The number	14	/ 2 = 7
25	50	50	of electrons	18	/ 2 = 9
36	72	72	per shell are	22	11
49	98	98	not proven	26	13
64	128	128	beyond the fourth level.	30	15

Table 2: Chart depicting one gram acceleration @ 1 dyne per. sec. per. sec.

time/seconds	0	1	2	3	4	5	6	7	8
centimeters per second	0	1	2	3	4	5	6	7	8
distance in centimeters	0	½	2	4½	8	12½	18	24½	32

Table 3: Timetable of a falling object			
Time in seconds	elevation from ground in feet	divided by 8	difference
0	256	/ 8 = 32	
1	240	/ 8 = 30	2
2	192	/ 8 = 24	6
3	112	/ 8 = 14	10
4	0		14

When these quotients revealed by the denominator 8 are added together cumulatively, the sums progress 2, 8, 18, 32, et cetera. The velocity of a falling object equals the number of seconds x 2 x 8 = feet per second. The total distance traveled equals the number of seconds^2 x 2 x 8 = feet. The distance traveled in the preceding second is found: [(#sec. x 4) - 2] x 8 = feet.

The timetable of a falling object is a re-creation of one found in the book *Evolution of Physics* by Einstein and Infeld (New York: Simon and Schuster, 1966), page 202; this pertains to Galileo's experiment from the Leaning Tower of Pisa. I have added the denominator eight in order to show the correspondence with electron shells. The difference between the numbers is also put in to enhance the feeling of coincidence. This book also contained reference to Galileo's experiments with falling objects in which it was determined that the weight is equivalent to the mass—which, Einstein noted, is quite remarkable when you think about it.

Buzz Aldrin's demonstration of Galileo's principle on the Moon shows that mass is equivalent to weight on the Moon also. Mass being equivalent in two different gravitational fields is handled in Newton's second law equation covering gravity, F = ma (force equals mass times acceleration), but it is still extremely coincidental that the two are equal. I have often read the phrase that is attributed to Einstein, "God doesn't play dice with the universe." My understanding of this phrase is that Einstein didn't believe in coincidences in physics.

One more interesting correlation is between these energies and the square root of two. The square root of two is approximately 1.414213562. Multiplying this number by two equals 2.828427125; this second number squared equals eight (see Table 4).

Table 4
$(\sqrt{2})^2=2$
$(2\sqrt{2})^2=8$
$(3\sqrt{2})^2=18$
$(4\sqrt{2})2=32$

This seems to be a function based on the square root of two. By now, everyone should be familiar with Mandelbrot's fractal geometric designs. When Mandelbrot's equations are graphed in color, some of them resemble scenes from nature. Consider that the square root of two might be the reason for the symmetry in the universe.

You can derive light from electricity and vice versa. Light and electricity have many similar attributes. Trying to link them as different configurations of the same phenomena is natural.

Michael Faraday performed an experiment in an effort to do just that. Faraday's reasoning was that if light is an electromagnetic wave, you should be able to deflect it with a magnet. This was tried on light passing through a slit with alternating magnet fields, and even through a prism, looking for any interference—all to no avail. Faraday did make the significant discovery that the plane of polarization of the light passing through the prism was altered.

Following Faraday's lead, Pieter Zeeman discovered a twinning of spectral lines using a diffraction grating and a strong, steady magnetic field. Figure 9 depicts the sodium spectral lines for simplicity because sodium only has two signature lines. It is a good monochromatic light source emitting only yellow light, which cannot be further divided into a spectrum by a prism. An oscillating magnetic field should interfere with light if it were a manifestation of the same phenomenon. The Zeeman effect is the only way light is affected by a magnetic field; this is explained by the spin of the electrons. Electrons found in pairs spin in opposite directions. The angular momentum of the electrons causes the light to veer off in two directions (see Figure 9).

After pondering the results of these experiments, it seems more difficult to reconcile these two phenomena as merely different frequencies of magnetic waves with the same sort of propagation through the ether. Light seems to possess only two dimensions, whereas an electromagnetic wave has three, both phenomena being exclusive of a time element.

Polarity of light can be caused by light passing through any one of a myriad of materials or simply being reflected off a puddle of water. The light looks the same as any other light until you try

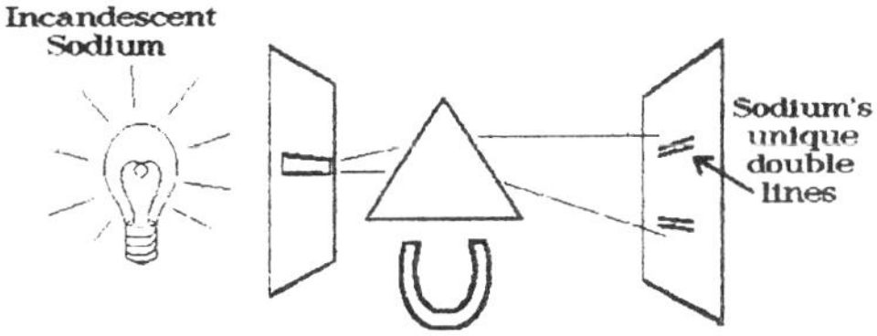

Figure 9

to pass it through an analyzing material. Polarity orientation of the light beam can be established using an analyzer.

There are also right-hand and left-hand crystals of an analyzing material. The light seems to pass through them in a screw-like manner. In 1848, Louis Pasteur noticed two crystals of a certain tartaric acid that were a mirror image of each other. After dissolving them in water, he discovered the polarity of light passing through was exactly opposite. One crystal passed light in a right-hand screw pattern, and the other was left, so you can observe two opposite photon paths within the same chemical.[1]

When photons are absorbed, the atoms expand. When photons are given off, the atom shrinks. This expanding-shrinking cycle would naturally synchronize itself to correspond with the frequency of the incoming light wave. This would also create a state of vibration in the material. Each material's molecular chain has a characteristic vibration rate, or harmonic, which is individual to that sort of material.

Have you ever looked at power lines on a cold, clear day and wondered why they seem to disappear at the point where they intersect? As the wind blows the lines, the point of intersection moves back and forth, and so does the illusion of transparency. When looking at two lines whose paths intersect at varying levels—for example, when one line is crossing the road at a lower and closer level, giving the perception of crossing—both lines appear to be bent at the point where they intersect. As you drive past in your car, this apparent bend travels with you. This gives the illusion the line has a wave passing down its length, keeping pace with your car (see Figure 10).

What could cause this? Light is a wave; by definition, this means it arrives at regular intervals. The atoms of the wire would be accepting photons from the sunlight. Each atom, when it accepts a photon, expands instantly. The atoms return to ground state almost as fast. This expanding-shrinking cycle would be amplified by the length of the wire. The end-to-end push of the

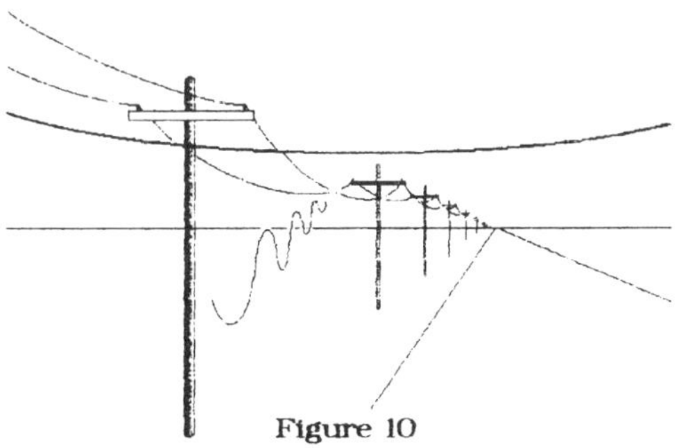

Figure 10

atoms causes an exaggerated vertical jump for the atoms of the line. Light passing the vertically vibrating wire becomes polarized by the movement of the wire. The light wave would now be perpendicular to the wire. Because the second, closer wire, is moving up and down, it acts as an analyzer. It will only allow vertical light waves to pass. I theorize that light travels through translucent materials in the same manner as it seems to pass through the parallel telephone wires. If we treat the passage of light as the quantum jumping of the atom from the path of the photon, then it would change how we look at other phenomena, including Brownian motion. When pollen is viewed under a microscope it is constantly in random motion. This was discovered by Robert Brown, hence the term *Brownian movement*.

Having the light pass through a narrow slit first greatly reduces the focal length needed for obtaining a spectrum from a prism. I theorize that the light striking the surface causes an expanding-contracting cycle for the gate material. This vibration filters out and reorients cross-polarity light waves, forcing those making the transition to stay within the limits imposed by the vibration of the side material. This light passing through the hole is restricted to one plane of its existence, thus polarizing it. The orientation of the slit is aligned parallel to the prism's sloped surface in order to display the spectrum in its largest form.

The idea I'm trying to convey is that, on an atomic scale, particles do not move in the conventional manner; rather, they jump from place to place without occupying any space in between positions. This is not an original idea; it was initiated by the math of Max Planck, and applied by Niels Bohr. Then it was put into words by Werner Heisenburg. Applying this concept to visible objects, such as the power-line example and hole diffraction pattern, seems unrealistic, but the explanation fits the observed problems. When light goes through two holes situated side by side, the material comprising these two gates is caused to vibrate. The rate of vibration is orchestrated by the rhythm of the light passing through versus the material's vibration rate. The light passing closest to the edge of the hole is deflected by the proximity of matter that is oscillating. The rate and amount of movement is also governed by the width of the material between the holes. The harmonic of these sets of variables regulates the intervals, allowing the light to hit consistently in the same spots and avoid the shadows.

This concept of the matter influencing the passage of light would account for the pictures being inverted in a simple hole camera. A hole camera is constructed using a cardboard box with a small hole punched in it, which will display an inverted view of scenery on the surface opposite the side containing the hole, and is sometimes used for viewing the Sun during a solar eclipse. The box's surface, with the hole, acts as an amplifier for the hole's movement. This causes the small hole to filter all but the single ray coming from the distant objects. This single ray will also be distorted; its amplitude will undergo changes caused by the surrounding surface vibrations. I also expect the light reaching the screen from a hole camera to be polarized with respect to the side it is deflected from (see Figure 11).

A spherical concave mirror can project an inverted image on a screen. These inverted images are referred to as *real images* because they appear to exist where the rays of light are focused. Right-side-up reflected images are called *virtual images* and cannot be projected onto a screen. Concentrating the light from a virtual image only brightens the spot it lands on and does not give any hint of the image projected.

The inverted image produced by a spherical concave mirror passes through a single point on its path to the screen, and that point is caused to resonate at the frequency controlled by all rays passing through that point. The rays are also distorted from real time, by the curvature of the mirror's surface, amplifying the wave vibration amplitude of these photons. I refer to the wave distortion taking place in real time; in the chapter on laser light I will be more specific on the time aspect of light as it relates to matter.

Light has two dimensions. The prism is the tool needed to

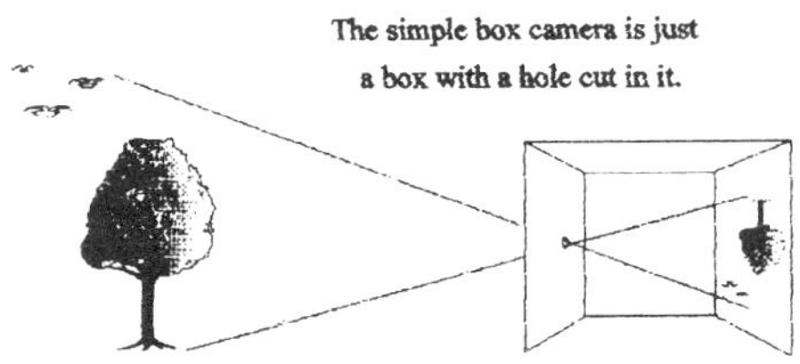

Figure 11

determine the height of a light wave. It seems reasonable that the wave height does not have to be equal to its wave length. The more radical turn, which is made by blue light, could be the result of the blue light wave being taller than the red. Try to imagine under what conditions this might happen (see Figure 12).

Hypothetically, it seems feasible that red light, separated from sunlight, was generated by an atom that is closer to the center of mass, which is under compression from its surroundings, and a slower time-rhythm from being at a more central location. Being made closer to the center of mass would slow its frequency, while the pressure from its surroundings would compress the atom and the light wave height coming from it. The blue sunlight would then be radiating from the larger surface atoms that are in a faster time zone. The conditions prevalent at the surface location producing the blue photon, having a faster time-rhythm with less compression, would cause the photon to be taller in height but more frequent in arrival.

Light from local sources also displays the same order of

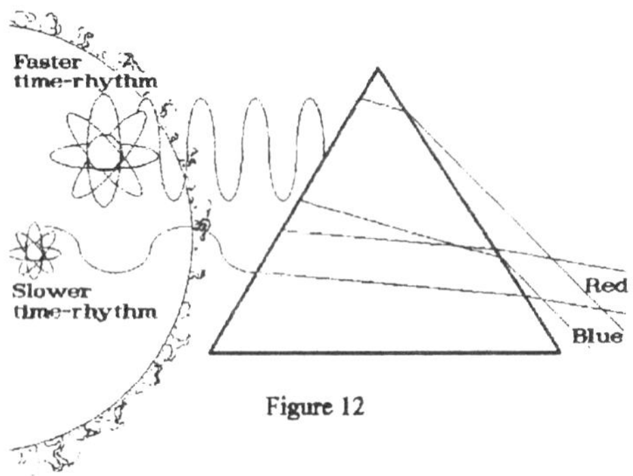

Figure 12

colors when separated by a prism. It could be that inside an atom there is more than one time-rhythm within its borders, and conditions exist to compress the red light wave's height.

This new line of reasoning would then cover the dual nature of light without it actually being both particle and wave. What we call space might be an exponentially less dense ether material conveying light's transmission. Even between galaxies, it is estimated there are three hydrogen atoms per cubic centimeter and we do have the solar wind within our solar system.

The reason these effects are not obvious is due to the ether, which is altering your perception of these events. Figure 12 gives a better understanding of how the principle works, and remains unseen to the observer of the refraction pattern for what it really is (see Figure 13).

The experiment is prepared by partly filling a fish tank with water containing a trace of fluorescein powder (fluorescein is a powder that sparkles, allowing the path of light to become visible). Then brine is added by carefully pouring it through a small tube extended to the bottom of the tank. This keeps the brine from mixing with the water. Now shine a narrow beam of light through the water with the fluorescein at the level that it meets the brine. The path of light appears straight to an observer

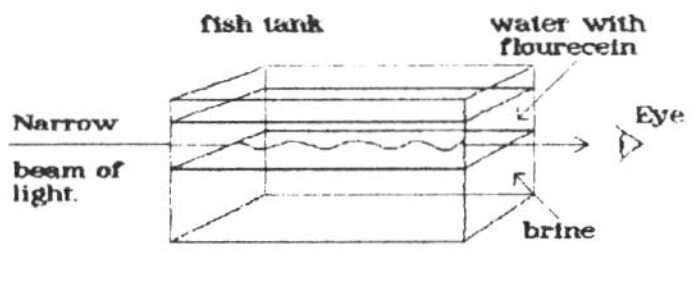

Figure 13

looking down the beam toward the source. If you move to a side view, perpendicular to the path of the light, the beam now appears to be wavy. Ordinary water with just the fluorescein added does not display this effect. The light is trapped between the two layers of water and the angle of penetration is too great to overcome; the path is disturbed by isolated thermal pocket differences.

With the new parameters set up, we can now start speculating how everything interacts. If we take a lesson from starlight, the effect might not be the pulling of light around the edge of the hole, but pushing the beam away in the circular diffraction pattern. Look at the shadow of your hand in sunlight. The shadow is composed of the initial shadow that is approximately the same size as your hand surrounded by a number of lighter shaded, well-defined outlines of the original.

There is an enigma found by blending colors. If you mix red, green, and blue, you obtain the color white. If you add red and green you make yellow, green and blue results in cyan; blue and red makes magenta. Red, green, and blue are grouped together under the label, "additive primary colors" whereas yellow, cyan, and magenta are known as the "subtractive primary colors." When all three subtractive primary colors are combined, they create the color black. Combining two of the subtractive primary colors can result in the display of one of the additive primary colors—for example, cyan with yellow makes green, yellow and magenta is red, magenta with cyan makes blue. It seems strange that combining the three subtractive primary colors results in black, when all the waves are present that result in making white. The resulting wave form depends on the order in which you combine them.

All matter is in a state of vibration at its molecule's particular characteristic rate, corresponding to the surrounding environment's energy level. Light energy adds to one of the two possible vibration dimensions of the vibrating matter, either side to side or expanding and contracting. This is the underlying relationship between light and matter.

Note

1. *The New Encyclopedia Britannica,* 15th ed., s. v. "colour."

Four

Gravity

An explanation for gravity has been very elusive. We also have an inherent desire to lump all forms of energy together under a grand unification theory. The existing gravity theories leave a scenario where the center of the Earth would have to be hollow. If matter pulled on matter, the center of the Earth would be a spot where the gravity pulling from both sides of the Earth would neutralize itself. This would cause the light elements to settle at middle Earth. Scientific consensus is that the Earth has a solid core (see Figure 14).

Present theories assume that space is drawn towards massive bodies. Particles would then be peeled from their assigned space

Figure 14

at the planet surface by electrostatic forces, thus creating the deceleration that we experience as gravity.

Before jumping to conclusions, we should consider what other possible options there might be. It is impossible to tell whether you're being stopped from falling by the surface or if the surface is accelerating toward you. At this point, it seems improbable that the Earth's surface could be accelerating toward you.

The Earth is spinning, and yet a man located at the equator would weigh the same if he stepped on the scale while visiting the north pole. Centripetal force equals mass times velocity per second squared, divided by the radius of the circle. Given the two different sets of conditions, there appears to be a discrepancy in either measuring weight at different latitudes on the globe, or with current theories.

In space, when the shuttle goes up to place a satellite into orbit, the shuttle and satellite can maintain orbits close to each other. When you tie a rope from the orbiter to the satellite, the satellite then attempts to gain altitude above the spacecraft. If you cut the rope, the satellite will jump seven times the length of the rope higher, while the shuttle drops to a slightly lower orbit.[1]The floating rope will align itself perpendicular to the Earth's surface. The rope's more massive portion will point toward the Earth and the lighter end toward space. Even the Moon acts in this manner, so that we see only the one side of it. Centripetal force appears to come back into play beyond the Van Allen belt, whereas below the Van Allen belt objects do not show this effect. (The Van Allen belt is an electrically charged region above the Earth's atmosphere that acts as a filter, eliminating most of the more harmful X rays; it also interferes with radio communications between spacecraft and ground control when the craft is passing through it.)

Think about it. All visibly discernible stars and planets are almost perfectly spherical when some should be disk-shaped. The Earth has a polar diameter of 7,900 miles compared to its equatorial diameter of 7,923 miles. The planet Jupiter, for example, is 1,300 times larger than Earth and is only 318 times as massive, being composed of mostly gaseous material without a solid core. It rotates in just under ten hours. Since it is made of a pliable material and has such a fast rotation rate, it seems as though Jupiter's ellipticity would be more pronounced than 6.5%. The equatorial diameter is 88,678 miles compared to its polar diameter of 82,928 miles. The surface is so malleable that it actually rotates at different speeds in higher latitudes. At the equator, its rate of rotation is nine hours, fifty minutes, and thirty seconds; in higher latitudes, it is nine hours, fifty-five minutes, forty seconds, and its magnetic field rotates in nine hours, fifty-five minutes, twenty-nine seconds. This multiple-layer rotation is probably responsible for the planet generating more heat than it takes in from the Sun. Using conventional arguments of physics, it would seem that the centripetal force should overwhelm the gravitational force and cause Jupiter to flatten out.

Here on Earth we have an isotropic distribution of minerals, which should result in noticeable gravitational imbalances. Gravity is not detectable between two objects situated side by side here on Earth; however, it manages to hold the Earth in orbit around the Sun. For a little perspective on that statement, it's like having a grain of rice orbit a golf ball over a distance of approximately a football field.

An experiment has been performed that showed an increase in gravity by positioning a massive lead sphere under one of two equal weights suspended on a balance beam. If the lead sphere were pulling on one of the weights on the scales, you should also be able to detect a decrease in gravity by positioning one on the next floor above the balance beam. There is no detectable lessening of the measured weight while positioning a heavy mass above the load.

Gravity could be the result of time being slower in the

presence of mass and with a compressed length measurement. This could make an atom lopsided in its emission of photons.

Time is pliable. If you have two synchronized clocks, and take one of them to the top floor of a ten-story building while leaving the other at the ground floor, the one on the top floor will, after a period of time, move ahead of the one on the ground floor. If they stay side by side for the same period of time, they remain synchronized. This experiment has been performed by several groups, I'm not sure who actually did it first (see Figure 15).

The theory of relativity tells us a moving clock experiences a slowing of its time-rhythm, so moving through the ether does have an effect on matter. Proof that time scales adjust is in the rosette orbit of Mercury. That also proves that the comparative motion of matter is controlled simply by the presence of the ether. Another cosmological example of varying time scales would be galactic rotation, as noted earlier.

As stated in the previous chapter, all matter is in a state of vibration. What we see as color is the rate of vibration of that matter. The photon has two dimensions. A photon itself could

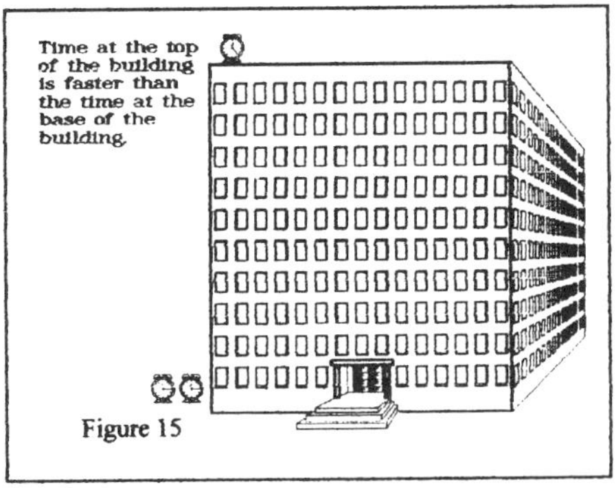

Figure 15

be the act of recognition that a new time-rhythm has been established between atoms. The shape of the photon could determine the degree of reaction to that information. This would result in a change of momentum for both atoms. The net results would effectively be the shift of a segment of space between the two particles. The amount of gravitational pull could be encoded in the height of the light wave as it relates to its length.

A massive object shrinks comparatively faster than lighter matter, and generates heat, as is observed with a collection of uranium. In this manner, the molten core of the Earth would generate heat without having a huge deposit of radioactive elements for the heat source, as proposed by Lord Rutherford. Without some source of energy, the Earth's core should have cooled eons ago. This would come about by having a slower time-rhythm making the photons for a longer period of time, thus producing long wave radiation.

From an atom's perspective, it would be receiving multiple short light waves from the side facing the lesser mass, and elongated light waves from the side facing the greater mass. The material surrounding the mass would be sending an abundance of shorter light waves toward the higher concentration of matter, displaying the characteristic appearance of a blue-shifted object moving toward the mass. This would squeeze the long-wave photons out of these atoms by putting the material under pressure. Radar is an example of a blue-shifted signal rebounding from the smaller-dimension space-scale of an approaching object, where a receding object will have its reflected signal red-shifted. I'm suggesting that the relative appearance determines the actions of the object (see Figure 16).

The atoms themselves would effectively be emitting short light waves toward the mass and long light waves toward space, giving the effect of traveling momentum toward the mass. Between two isolated atoms, one of which is in a higher energy state, the movement of space would be from the sender of the photon

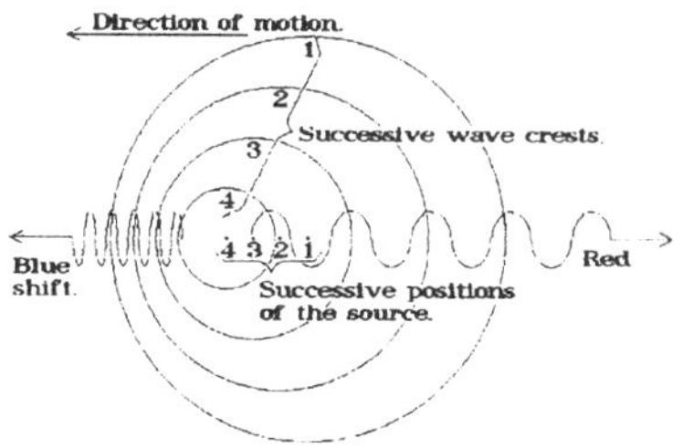

Figure 16

to the receiver. The sender, devoid of the added energy, now has a shorter length measurement, so that the distance between the two atoms now seems to have grown. Conversely, the receiving atom's length measurement would now be longer. With the atom's newly acquired length scale comes a slower time pulse relative to the length of the absorbed photon (see Figure 17).

This description of gravity as photons distorted by differing time scales eliminates the problem of the natural expectation of having a hollow Earth, which would be the result of having particles pull on particles.

This is where I differ from current theories. I'm postulating that the star, planet, or whatever, is manufacturing space, but the momentum is inward due to differences in relative time between the layers of matter. The rhythms would be slower as you move closer to the center of the mass.

In the past, the circular orbit of planets has been described as a geodesic—a straight line through curved space. I equated a photon to the movement of space away from the object emitting the light. Figure 18 shows how light passing close to a star would be diverted by segments of space moving in opposite directions. I go on to show how this continuing process would bend starlight

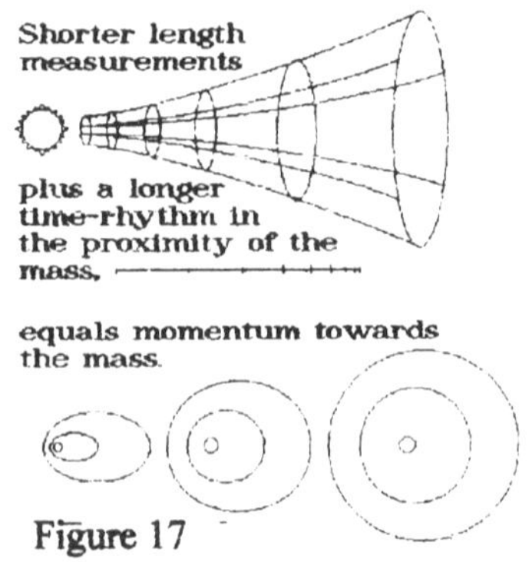

Figure 17

that is passing close to the surface of a star. Consider each elevator as a radiant segment of space emanating from the star. With the advent of my theory, I have to change the terminology from *traveling through curved space* to *redirecting the momentum of traveling objects through a photon exchange* (see Figure 18).

Anytime two particles are present, a mutual distortion of their photon emission occurs. A moving object would have its velocity factored into its slight realigning due to the relative time difference (satellites do fall from orbit). An atom that is broadsided by a photon traveling perpendicular to its path has a greater influence on the trailing side that is emitting the larger segments of space, as though it had different time scales within its breadth. It was once thought that when a satellite was placed into orbit, there it would remain forever. That is not the case. Several satellites have recently had their orbits decay and have burned up in the Earth's atmosphere.

The momentum of the object is decided by the amount of photon distortion calculated through the relativity equation experienced by the passing of coordinate systems. The effect obviously travels through solid matter. This allows you to feel the sensation of acceleration as in an elevator. The distorted photon

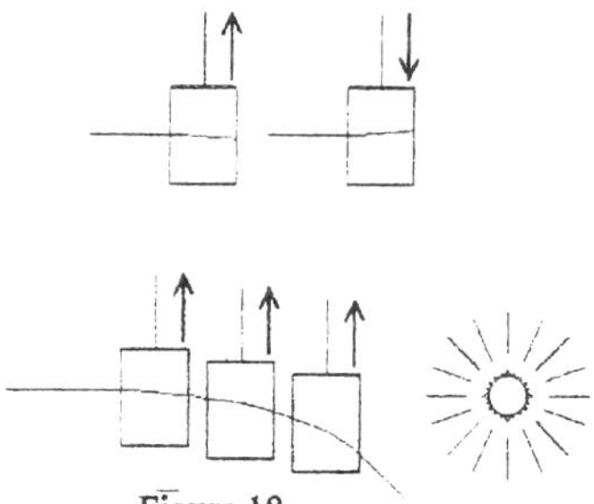

Figure 18

finds matter translucent. Long-wave energy always seems to have greater penetration.

The closer proximity to the mass emulates an object moving by being intersected with greater number of long-wave photons emanating from that mass. This greater concentration of energy is equivalent to faster velocities.

What I'm suggesting is that it all boils down to an adjustment of a time relationship between objects. According to Einstein, there is no difference between gravity and centripetal force: acceleration or deceleration, which is merely a redirecting of momentum. Consider centripetal force: you have the difference of comparative photon lengths present, between the outside material and the rotating disk. The time relation between the disk and its surroundings is distorted by the two moving coordinate systems. When I was a child playing in the sandbox, I remember pouring sand past the spinning wheel of an overturned tricycle. I wondered why the sand was attracted to the rubber wheel. Several explanations for this present themselves. The explanation I settled on was that the air carried by friction with the spokes was forced outward by centripetal force. Then the sand was drawn back into the tire by an eddy current of air. Now

I feel that the photon length-to-width ratio differences cause the attraction. Neither the rubber wheel nor the sand have magnetic properties. Sand, as silicone, is the main ingredient of glass, and a glass rod rubbed on either rubber or fur develops a positive charge. The observed attraction occurs without contact being made. It would seem that proximity and movement govern the changes without the obvious flux lines of a magnetic field. I am saying that when you have two coordinate systems moving relative to each other, an attraction between the objects contained therein is created. A near miss in the vacuum of space should cause a turbulence for the target.

The rotation of the Earth should have you feeling lighter at the equator due to the centripetal force, but experimental results do not conform to our expectations. The Earth is approximately 24,902 miles in circumference and rotates once every twenty-four hours, so the speed of a person near the equator is roughly a thousand miles per hour. Einstein must have really dwelled on this point when he stated that a man spinning in an isolated universe wouldn't feel the spin if he were alone in that universe.

It is difficult to make the transition from the rock-solid logic of Newton's ideas to the elusive proposals of Einstein. Discrepancies in the way gravity is not detectible between massive objects side by side, but acts over long distances between the Earth and Sun, beg for new answers. If you relinquish preconceived notions and accept the proof of an atom bomb, the changes inherent to the relationship of energy exchange between objects brought on by relativity do make sense for describing attraction between bodies.

In the theory that I'm expounding, the man would be adjusting his time scale to the matter that is moving with him directly under his position. With current ideas, the force of gravity would have to become stronger with the increase of velocity in order to ignore centripetal force as a factor. In the planning for building skyscrapers, they completely ignore the consideration of centripetal force in all latitudes.

In order to accelerate or decelerate, energy is consumed. This means matter undergoes a transformation into photons. Friction is inherent in any object passing another. Even water being stirred shows an increase in temperature. The atoms, when passing each other, distort each other's electron-shells, and when they snap back they emit a photon so that movement is directly converted into energy. This was first experimentally demonstrated by James P. Joule, a brewer in England in the mid–eighteenth century, who had the facilities to rig up a paddle device in an isolated vat and measure the temperature rise per distance of a falling weight. His conclusion was that it takes 772 pounds falling one foot to raise the temperature of one pound of water from 55 degrees to 56 degrees Fahrenheit.[2]

The thrust of a jet engine is an aspect to consider as a violation of Newton's law of motion: For every action there is an equal and opposite reaction. By strictly adhering to Newton's axiom, the jet engine is capable of generating more thrust than deemed possible within these rules. The amount of thrust created exceeds the energy calculated by multiplying the mass times the velocity of the escaping gases. In order to allow for these differences, it was decided that the thrust is caused by the gas pressure on the remaining walls of the combustion chamber. The pressure is controlled by a confinement of the flow of hot gases at the nozzle. This may be a hint that the comparison of rhythms of an atom to its immediate surroundings is responsible for the pressure it exerts. Remember your grade-school science class, when the teacher explained the flight of the balloon? The explanation was that the flight of the balloon is the result of air pressure on walls that don't have a hole releasing air. The thought that something is missing must bother someone other than myself. What the jet engine demonstrates is that under pressure, particles assume a slower time-rhythm, which would have to be more massive than their counterparts that have been released from pressure, in order to account for this discrepancy. Also, current aerodynamic theories can't explain the flight of a bumblebee.

Light holds the greatest prospect for an answer to the gravity puzzle because of the photometer. The photometer demonstrates that light can be directly converted into motion.

The photometer is constructed using a belljar that has been evacuated and has plaques that are painted black on one side and white on the other which are suspended to facilitate free rotation. The plaques rotate rapidly from the black side toward the white side when exposed to light (see Figure 19).

The problem of having a long-wave photon represent gravity is also caused by this experiment, which demonstrates that a long-wave photon moves away from the stationary surroundings, while a short-wave side moves toward them. Gravity would necessarily have to be a long-wave form of energy from its closer proximity to a greater concentration of mass and its inherent slower time-rhythm. This inconsistency seems to eliminate the

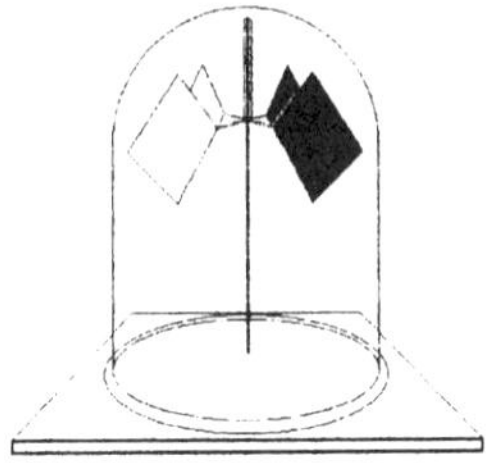

Figure 19

possibility of the two forms of energy (light and gravity) having any shared characteristics.

After carefully considering the photon, it should have two dimensions used in determining its potential, not just the frequency of its arrival. The atoms of the interior of the vacated belljar would be in a faster time-rhythm due to the vacuum. The black side under these conditions would be generating taller photons combined with the longer wavelengths. The sum of the two photon dimensions is greater than normal photon emissions of local matter including the ones found coming from the white side of the plaques. The photon emissions from the black side then repel from local matter.

Using the possibility of a light-wave height difference, consider the photons from the top of an airplane wing, which has the faster moving flow of air when in flight. The Lorentz Transformation theory tells us that the visual image of the wing will have shrunk in the direction of its motion. Einstein's relativity also states that it will possess a slower time-rhythm by virtue of moving. These photons will be red-shifted in length, they will also be skinnier in width (see Figure 20).

The photons from the wing, as seen from the top, would have the same appearance as if they were coming from an object that is red-shifted by moving rapidly away (elongated without gaining any height). We might then consider that by warping a photon's length-to-width ratio we can create a condition simulating gravitational pull.

To consider the lift of a wing using classical physics, a broad wing would logically experience the greater lift due to the larger surface area for the differences of air pressure to act upon. This is not the case: Long, narrow wings show more efficiency. A helicopter was considered an impossibility in the 1930s. The top of the wing is making contact with more atoms; like the Venturi effect in a carburetor, which experiences the greatest reduction of pressure at the spot that has the highest concentration of

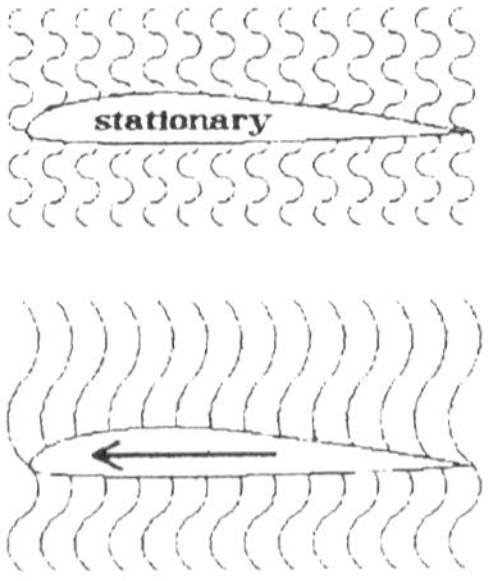

Figure 20

molecules, speed is the governing factor. The air behind the wing reveals the absorbed energy when viewed through an infrared scope. I don't believe the currents of air behind the wing could be responsible for the lift at a distance without considering it as gravitational pull.

A sailboat can achieve faster speeds than the wind pushing it. The curved front surface of the sail actually pulls the boat forward—it's not just a matter of the air being deflected toward the stern by the curved sail.

Ultraviolet light possesses greater energy for the purpose of knocking out electrons from their atoms. The compressing of the wave height seems to give the photon greater energy to be absorbed, and the elongation conveys higher temperatures. Planck's formula $E = hf$ used to determine the energy of a light wave may only represent half of the picture. The height of the wave, or lack thereof, could be a multiplier of the energy to determine the photon's gravitational pull.

The fast quarter-turn change in flight orientation of a Frisbee, followed by the stable gyroscope-type flight when the new photon exchange rate and time relationship has been established, suggests to me that possibly some sort of electrical-style change occurs (electrons have been found to exit from a hot oven in a regular order, the second perpendicular to the first, and the third perpendicular to the preceding two). Even though the Frisbee wobbles, I don't consider that it might be due to an unstable gyro-motion because it levels out to a smooth gliding action. The time relation overrides the physical attributes of the object, causing the object to orient itself in the manner that is dictated by the new time relation to its surroundings. The new time relationship caused by the rotation allows the Frisbee to hang for abnormally long periods in one spot, imitating the unexplainable slowing of a comet at perihelion. This would also add a new perspective to the concept of drag on an airplane's wing. The difference of time-rhythms causes the electrons to appear to have established a new equilibrium (see Figure 21).

Consider how this different way of looking at the photon exchange could be used to explain the flight of a Frisbee. A Frisbee overcomes a gyroscopic influence in order to change from the vertical position which it leaves your hand, then angles upward at the end of its flight at a point where its rotation and speed have almost ceased. (I'm referring to the original model of the Frisbee. The newer Frisbee models have a heavy plastic ring in them to gyroscopically stabilize their flight.)

The electrons residing on the top side of the Frisbee that is spinning toward the direction of motion would occupy the side having the slowest time-rhythm. This would distort the photons emitted from these electrons by shrinking them in the direction of their motion, and elongate them perpendicularly to that motion. At this point, the Frisbee makes a quarter turn on its plane of travel (see Figure 22).

Another example illustrating this point would be the physics experiment demonstrating torque: Have a person standing on a platform that's capable of rotating freely. The person then holds a spinning bicycle wheel by a handle, as the orientation of the wheel is changed from spinning perpendicular to the ground to

Figure 21

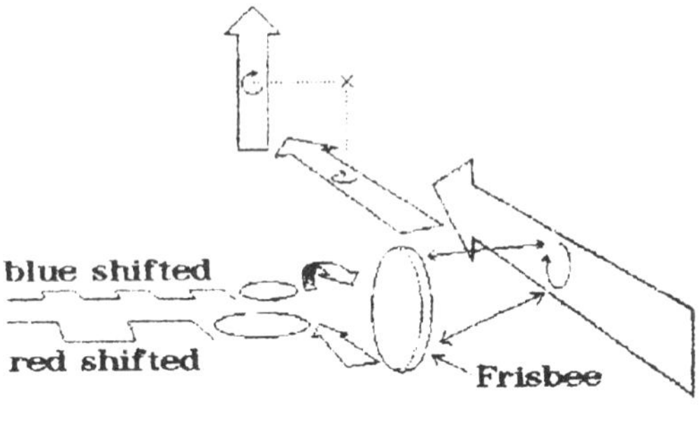

Figure 22

spinning horizontal to the ground, the person holding the wheel will start rotating so that the direction of the motion of the wheel which is away from the person coincides with the direction opposite his rotation, as though the wheel were getting traction on the surrounding air. If the wheel is turned over, the direction of rotation of the person will reverse.

The ice pail experiment by Michael Faraday demonstrates the difference in time it takes for the two opposite electrical charges to dissipate. A negative charge acquired through induction distributes itself evenly over all surfaces of an object. When it is grounded, it equalizes its potential to the new frame of reference at near the speed of light.

In the ice pail experiment, we are dealing with the positive charge. The positive charge is introduced by induction from a positively charged pith ball. This leaves a positive charge on the outside of the pail and a negative charge on the inside of it. The electron-type charge, being less massive, more readily moves and can then be drawn away, leaving the slower-moving positive charge behind. If the pith ball used to extract the charge is left in contact with the pail long enough, both opposite charges will slowly equalize to their frame of reference (see Figure 23).

The entire top surface of the Frisbee then becomes populated with elongated electrons, creating a static charge that is slow to dissipate. In this second stage, the interaction of the charge with the air allows it to hang in slow motion to its surroundings.

In the third stage of flight orientation, the slow-moving positive charge creeps over the lip of the Frisbee, distorting the photons from the protected faster time-rhythm bottom, stretching them horizontal to the bottom surface. The photon appearance governs the action of the Frisbee, pulling the blue-shifted bottom toward the direction of motion, so that the motion imitates the photon appearance (see Figure 24).

A small stream of water, flowing past a balloon that has been

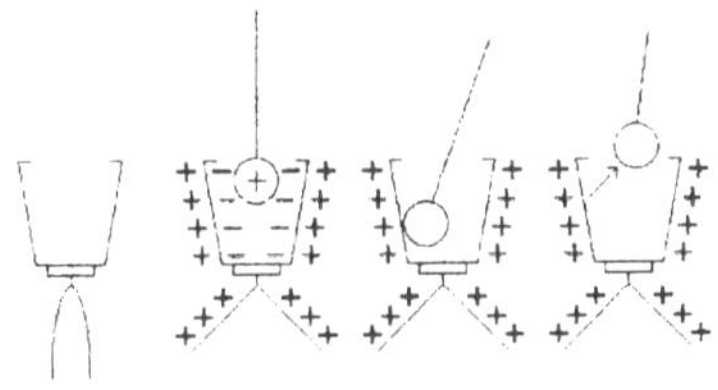

Figure 23

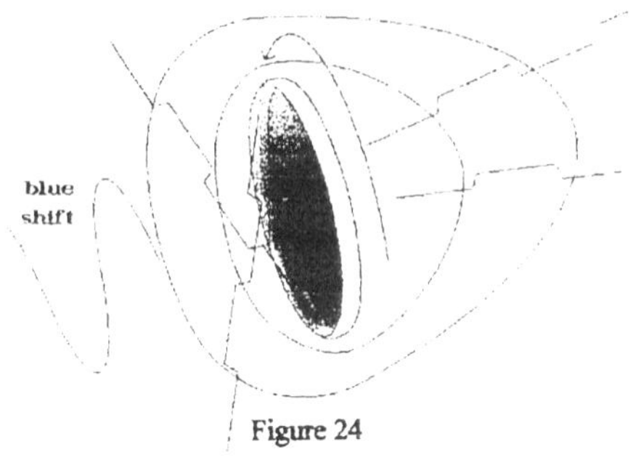

Figure 24

charged by rubbing it in your hair, will be attracted by the balloon. This observed attraction also contrasts with a wire conducting a current being repelled by a moving magnetic field. The two are both conductors. The charges are opposite, but the wire is repelled at right angles to a magnetic pole, whereas the water is pulled directly in, toward the balloon.

Following this trend of thought, the electrons that are distorted vertically to the surface might represent the positive charge causing a linear photon distortion, resulting in an attraction between bodies (see Figure 25).

Similarly, an electrically neutral cloud passing over the ground develops a positive charge by having its view of the photons it is receiving narrowed. The ground is then negatively charged by having an electrical charge moving in its vicinity.

Newtonian physics really has a dilemma explaining why a comet's tail is repelled by the Sun. There are many problems with the current model for comets. For instance, how do they replenish their gaseous material, which is depleted by the solar wind on each consecutive journey toward the Sun?

These following ideas may sound farfetched, but the action of a comet's tail requires a radical explanation. One of my purposes for this book is to put forward some innovative explanations.

The velocity of a comet slows to a walk at perihelion, then it catapults away at a million kilometers an hour. This act is contrary to all our present physics theories.[3] The planet Mercury also slows on its approach to the Sun, then resumes its normal speed at the apogee of its orbit. This change, however, is not as flagrant as that of a comet. From this clue, I assign the comet its faster time-rhythm (see Figure 26).

The tail of a comet stands straight out for millions of miles, always staying away from the Sun, and grows longer as it approaches perihelion. The typical comet orbit is not on the same plane as are the planets. Comets cut through the planets' orbits,

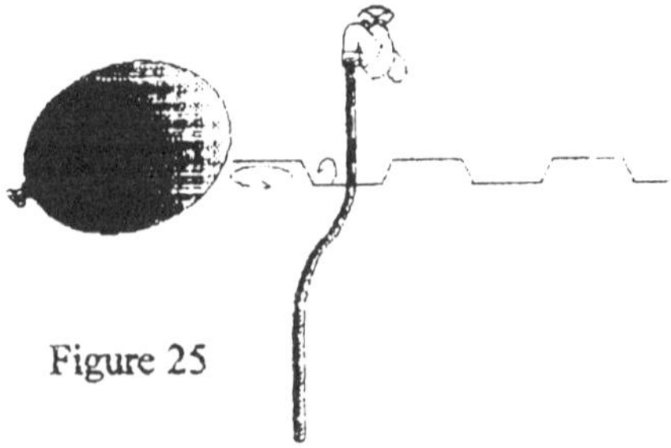

Figure 25

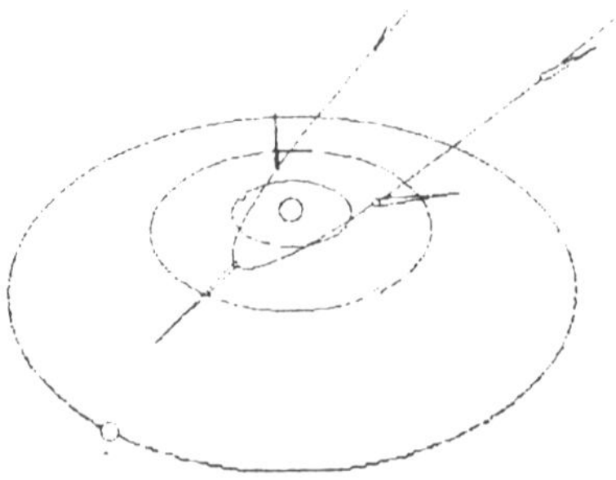

Figure 26

and, as Halley's comet did in 1910, across the Earth's gravitational field. I reject the explanation that the solar wind blows the tail away, because the solar wind doesn't affect our planet's atmosphere in a similar manner. Instead, the Sun's gravity pulls the Earth's atmosphere toward it, in a tidal effect.

On Earth, our atoms are limited to eight electrons in the outer shell at ground state. These electrons set the rules involved in chemical bonding. Imagine the red spot on Jupiter, which has survived for centuries with only the lighter elements available, and is in an unbelievable state of turmoil. This next thought is from the differences that could be encountered at faster time-rhythms. We do not have a model for the chemical-style bond that I'm proposing. Atoms may use many more electrons in their bonding, or valence shell, in a faster time-rhythm. Consider that an atomic implosion rips the electrons from our time-rhythm to disrupt chemical bonding as we know it. If this is the case, then the comets would not have to divest themselves of matter on each approach of the Sun. I'm speculating that the elements composing a comet, including the tail, are tied together as a unit connected by a chemical bond beyond any found in our time-scale vicinity. The tail portions, which are at a stable thermal state and reluctant to shrink, help to drag the head back out to an area closer to its natural time-rhythm, like being tethered to a rubber band. Comets require something outside our normal experience, so stretch your imagination. I readily admit these ideas are stretching the limits of believability.

When the planet Mercury enters the slower rhythm of the Sun, it may also enter an area that only allows two electrons in the valence shell. The giant gas planets such as Jupiter, on the other hand, may have eighteen electrons in their valence-shell, allowing a chemical bonding that holds up under extreme conditions. A comet's tail elasticity might be caused by electron shells that are well beyond that. Size, speed, and the elements present all play a part in this assessment. I would like to see a satellite launching a smaller satellite using the rope-tether method while orbiting Jupiter, in order to determine if the distance the satellite rises is seventeen times the length of the rope, while the more massive satellite drops slightly. The recent probe that landed on Mars found that chemical reactions in experiments do not perform as expected. This may be a hint that Mars occupies a spot on the border of a faster time-rhythm.

The rock samples brought back from the Moon indicate that it is at least a billion years older than Earth and has an independent chemical composition. This information discredits the theory that everything in our solar system condensed from the same gas cloud at roughly the same time. This would also cause a mystery as to how the Moon attained a near perfect orbit around the Earth. The lunar exploration also found magnetic hot spots. These locations must have experienced extreme electrical fluctuations to sustain a magnetic disturbance.

Astronomer Michael Mendillo of Boston University has determined that the Moon's faint sodium and potassium atmosphere acts like a comet's tail. The sunward edge has sodium extending to about 7,000 kilometers; the shadow side is fainter but reaches 21,000 kilometers.[4]

Try to imagine an atom's vibration state. There are two possible vibration qualities available. The atom can either vibrate side to side or in an expanding-contracting fashion. Now imagine what kind of properties these vibration qualities might impart to matter. With a faster time-rhythm and a greater side-to-side vibration, there may develop more locations at which an electron can establish itself in the valence shell. I will explain these ideas in more detail in the chapter on electrons. With eighteen electrons in the valence shell, there would be the possibility of plus or minus nine electrons for chemical bonding.

It has been suggested that comet debris may have been responsible for the Tunguska fireball. This explosion occurred over the Taiga forest at Tunguska in a remote region of Siberia in 1908. This had all the earmarks of an atomic implosion (an implosion releases more energy than an explosion), judging from the trees standing straight up at ground zero, without distributing any telltale radioactive fallout. This might be an example of what direct contact between matter of a faster time-rhythm from deep space, coupled with high entry velocity, can accomplish. An atomic structure collapse of the matter from the faster time-rhythm occurs. I will write more on this subject when I discuss nuclear triggering devices.

The Earth is rotating in a positively charged solar wind, causing the negative charge to flow at right angles exiting at the south pole. The magnetic pole changes polarity approximately every 50,000 years.

On a tangential thought, the number of intersecting long-wave photons that have a squashed height might cause the bone degeneration that has been documented by astronauts that have spent long periods of time in space. Even though an attempt has been made to regulate the astronauts' body temperature, I doubt that long-wave radiation has been thoroughly considered. Calcium links together in the bones in a specific manner. In the absence of the long-wave energy, calcium molecules might realign based on a different electron shell. Similarly, glass placed in water changes the glass's properties. Under six inches of water, you can nibble-cut glass with scissors. Under ten feet of water, it becomes impossible to break ordinary window glass with your bare fist.

The distance of the planets from the Sun has a relationship with Bode's law. Bode's law is a set of numbers that coincides with all of the planets' orbits at multiples of that set. Bode's law also predicted the presence of the asteroid belt, which wasn't known about at the time he suggested it, causing the search for their missing mass. A number of the planets have a quirk in their rotation. Some have retrograde rotation. Saturn's poles are tilted so much that we can observe its rings' flat surfaces at times, instead of just looking at their edge. The poles of Uranus are aimed directly toward the Sun, so that one pole enjoys a 42-year daylight period before slipping into an equally long night.

The formation of the planets probably has a connection with the vibration of the solar system. Recent studies indicate that the interior of the Sun vibrates. On a related topic, the Japanese who are studying the control of smokestack emissions through the use of sound waves transmitted into the smoke by means of a transducer found that vibration caused the smoke that would normally leave as minute particles to coalesce into masses 20,000 times larger than the original particles. These conglomerated particles then deposit themselves on the sides of the chamber.

If you take a flat disk and put it on a rotating table—say, a record turntable with music vibrating it—then sprinkle powder on the disk, the powder will redistribute itself into concentric rings with lines connecting these rings, resembling a spiderweb. This spiderweb type of pattern continues to the outer edge of the disk. The particle density diminishes as the dust travels outward. The high-velocity impact of a bullet with a car windshield displays this spiderweb idiosyncrasy. A number of other energy phenomena share this circular form. I watched a 1930 vintage film clip in which they were studying the concussion from high explosives. The destructive power seemed to bounce. An area immediately surrounding the explosion would be left untouched, this would alternate with destroyed regions in widening circles.

The Hubble space telescope does a slight Brownian motion kind of twist every time it comes out of the shadow of the Earth. If this were due to the combined gravitational field adjustment, it would seem as though it would develop gradually through the entire shadowside trip. Instead, the jump occurs with the appearance of light.

Researchers Michael M. Burns, Jean-Marc Fournier, and Jene A. Golovchenko found while experimenting with laser light that when they focused from two to five lasers through the bottom of a glass box onto micron-sized plastic spheres immersed in water, these low-power lasers created interference patterns. These interference patterns, in turn, caused the plastic spheres to rearrange themselves in a manner so that they avoided the dark patches and tended to stay within the lit areas. Two lasers lined up the spheres in rows; three can create hexagon patterns. With five beams, the result is a quasicrystal arrangement. The pentagon-shaped units were uniform in shape, but strewn together in an irregular manner. For this new bonding, they coined the term *optical matter*. The only bonding material involved is the light. The particles congregate where the greatest energy activity is occurring.[5]

Notes

1. Encyclopedia Britannica, s.v. "Science and the Future," 1991.
2. Albert Einstein and Leopold Infeld, *The Evolution of Physics*.
3. David A. Seargent, *Comets: Vagabonds of Space* (Doubleday, 1982).
4. Ivars Peterson, "Tracing the Glow of a Lunar Trail," *Science News*, June 8, 1991.
5. Russell Ruther, "Light Motif Binding Spheres with Light Forms Curious Crystals," *Scientific American*, November 1990.

Five
Laser

Laser light has three distinguishing properties: (1) It is nearly monochromatic, more so than normal light, even after normal light has been separated into its constituent colors by a prism; (2) It has coherence, with all waves as close as possible to being in-phase with one another; (3) it is directional, with the light produced having a narrow cone of divergence from its source.

The properties of laser light are remarkable. Looking at a white index card illuminated by both a laser beam and the ambient light of a well-lit room, the spot struck by a HeNe (helium neon) laser beam has a granular appearance. The spot illuminated by the beam is comprised of both bright red and deep black sections, of such size as to be easily discernable as separate entities. Even in the presence of the room light, the black portion is deep black. The most amazing aspect of laser light is that its path is visible without having particles present to bounce off of. Monochromatic light from a sodium light source is not visible, nor is light from the Sun, even close to its surface. Any other type of light requires a solid surface to bounce off from in order to become visible.

Light from a near source (referred to as an extended source of light) is treated differently than light from a far source (point source of light) within the human eye. The eye discerns a different characteristic depending on distance, and focuses light from a far source onto a single point of the retina.

Laser light, from a source nearby, is somehow confused with

a point source of light by the eye and is focused onto a single point, making it more dangerous for potential eye damage.[1] As an expanding photon sphere travels away from a single point, the wave front would appear as a flat plane at a point far enough away from the source. Light from a flat plane is treated the same as laser light by the human eye.

To cause matter to instantly give off heat energy is a simple matter of putting it under pressure, as in the compression stroke of a diesel engine or the compression side of a refrigeration unit. (The refrigerator was the first working laser; it gives off light [heat] through stimulated emission.) Let's take a simple-to-construct laser into consideration in order to simplify our understanding of how a laser might work. A good example would be the tunable dye laser designed by J. R. Lankard of IBM. I'm only concerned with the mechanics of it. If you want a complete schematic of the electronics, you will find it in the book *Light and Its Uses*, published by *Scientific American* (page 24). Another very good book on laser theory would be *Lasers, the Light Fantastic*, by Clayton L. Hallmark (Blue Ridge Summit, PA: TAB, 1979).

The amplifying medium is a dilute solution of one of three suggested dyes in a methanol-alcohol solution that is tunable to give a full range of the visible spectrum. Methanol-alcohol plays no part in the laser action; its only role is to suspend the dye. This solution is kept moving through a tube to prevent heating. Heating causes a distortion in the density of the column of dye. Distortion causes random differences of the optical refraction. Heat distortion doesn't present a problem due to the heating effect lagging behind the fluorescence by several millionths of a second. Only a small portion of the absorbed energy is emitted as coherent light.

The flash lamp consumes thirty-five megawatts of power and persists about two-millionths of a second. The capacitor is designed to discharge in that brief interval—ordinary capacitors are too slow. The bright flash is absorbed by an active medium and the photons caught between the end mirrors stimulate further coherent emissions, eventually escaping through the partially silvered feedback mirror. Light is synchronized in phase, by reverberating between mirrors, until it resonates at the frequency natural to the cavity.

Both flash tube and dye tube are enclosed inside a polished aluminum tube that is purposely squeezed slightly out of round. The need for the tube to be oval brings my thoughts back to the hyperbolically curved, reflective surface that is used to enhance the circular-diffraction pattern. With a flat mirror, or just a light source, the circular diffraction pattern is greatly diminished. Also, light emitted from this configuration illuminates objects brighter than light from any other method used to concentrate light from a single source.

On a tangential thought in a related field, in the Jacob's ladder experiment the distance between the two uninsulated wires gradually increases. The high-voltage electrical spark that starts by jumping the narrowest point will stretch to accommodate the gradual increase of distance between the wires. The voltage increases in small increments as the spark rises between the wires. The experiment will not work with a wide angle between the wires.

The span of the circular photon sphere reflected from the hyperbolically curved reflective surface gradually accelerates away from its counterpart as the light wave progresses through its expansion. All of the rays exiting remain parallel to the central axis of the surface, which accentuates its directional aspect. Possibly the photon can be gradually persuaded to represent a photon produced in a faster time scale. At a location with a faster time-rhythm, the height vibration characteristic would be measured over a greater surface plane area than that from a slower time scale. A photon created under these conditions would be a reflected wave aberration, representing a faster time-rhythm photon, which is relinquished by the receiving atom much more rapidly, accounting for the enhanced illumination property of matter struck by such a beam (see Figure 27).

Photons have no regard for time. Everything we observe, including light, has to stay within the limitation of light-speed. A condition for matter is to never exceed light-speed in vacuum, as measured using any time/length scale. The time scale is adjustable, though, depending on mass density, which can distort the perception of the wavelength of an incoming signal by the receiver. A light wave's length to width ratio will affect the time scale vibration rate of any matter that it encounters, and vice versa. When laser light passes through a small hole, its plane surface is restricted; the laser light loses its narrow cone of divergence after passing through an aperture. It seems possible that the divergence of the beam would be a product of the time scale it represents after being affected by the confinement within another time scale.

A physicist named John Clauser realized after reading a mathematical proposition put forward by John Steward Bell that a test of the two-photon puzzle was possible.[2] In his experiment, Clauser was assisted by Stewart Freedman. They found a way of

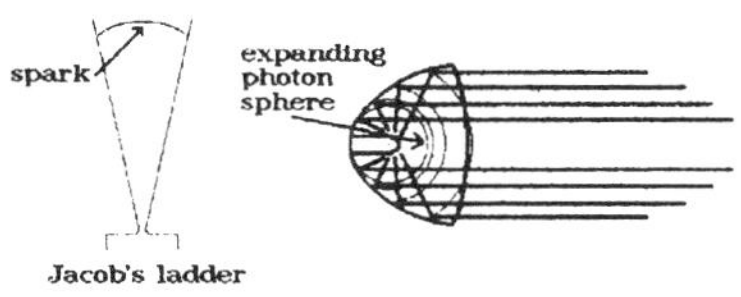

Figure 27

firing photons in opposite directions and selectively changing their polarization. The outcome was clear: A change in one photon affected the polarization of the other. This experiment raises more questions than it answers. What type of signal can overtake a photon to notify it of its counterpart's new polarity? Does the entire sphere of the expanding photon undergo a polarity change at the same instant?

I propose that the photon from the hyperbolically curved mirror is an inseparable unit that includes all of its reflected portions. When this photon is absorbed by the electron as a shell jump, it is a jump native to a faster time scale. This photon has an enlarged length-to-width ratio, which exceeds the normal ratio at the pressure of the dye, and is squeezed back out almost instantly. Consequences of absorbing this enlarged photon would be similar to what happens when Freon is compressed in a refrigeration unit. The result: The dye receiving the photon emits all of the acquired energy instantly, in a coherent fashion. The feedback mirror mechanism strengthens the uniform phase characteristic throughout the active medium. This creates a completely independent color from the initial flash, and the color is chosen by adjusting the distance between the end mirrors. Moving the mirror closer causes a shift toward the blue end of the spectrum in the resulting light beam. As the mirrors move farther apart, the light wave lengthens. This also depends on the limits imposed by the chemical's range of fluorescence (see Figure 28).

The idea that laser light is made as a result of an artificially faster time-rhythm leads to the thought that it may be flatter in its wave-front characteristic, causing a temporal displacement perpendicular to its plane of travel. What you observe when you look at a visible object is its time displacement, as it pertains to an atom's expanding-shrinking cycle. When looking at laser light, you're observing this same displacement offset for a perpendicular direction from the light beam (see Figure 29).

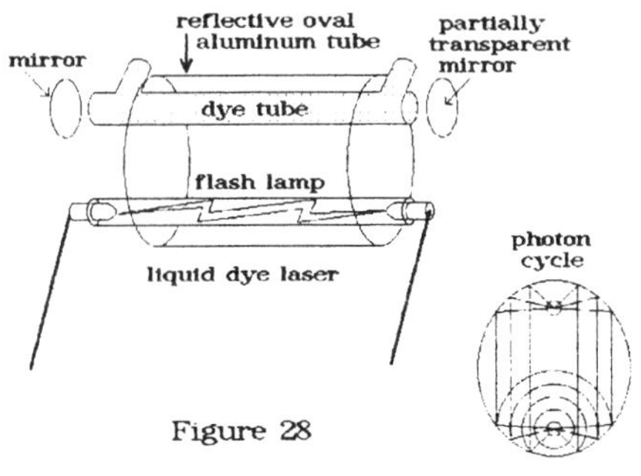

Figure 28

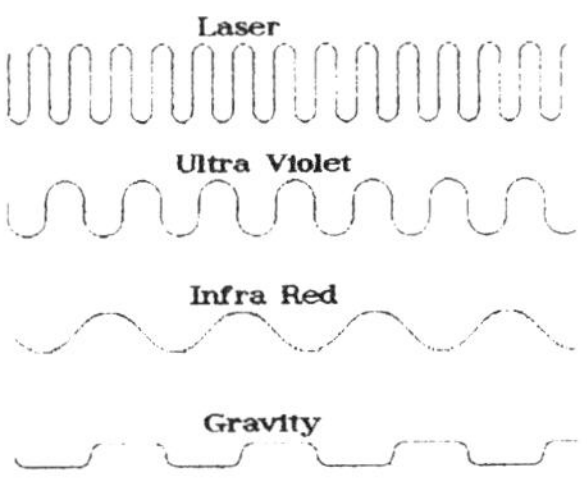

Figure 29

A light wave is many times larger than an atom, so that the wave amplification displacement would be measured by how many atoms in a single direction are affected in a single jump. The straightness of the wave line could also be a factor in its intensity. The abrupt change in output of a laser mode coincides with its power.

A maser waveguide was initiated by the knowledge that two antennae, side by side, intensify the portion of a radio signal that is emitted perpendicular to a line between the two. There is no signal emitted in a direction from the top of the antenna. The emitted signal from between the two antennae would be flattened from the signal of the opposite antenna. The signal emitted in a line parallel with the antennae would be still be rounded in their flux-line pattern. From these initial considerations came the wave guides that promoted standing waves within an amplification cavity.

A faster time-rhythm has an inherent longer length-scale for its side-to-side vibration characteristic, which is also a by-product of light reflected from a hyperbolically curved mirror surface. The synchronous disbursement of reflected light from a hyperbolically curved mirror enhances the circular diffraction pattern of a small hole by not being allowed to become a random photon of the surface material. The fast time-rhythm photon causes the side material on the surface around the hole to perform greater gyrations, displaying an interference spectrum between time scales. The multiple circles would be secondary harmonics developed within the conflict between the larger jumps, and the characteristic smaller ones found in the surface material vibration rate.

Figure 30 represents three colors of laser light passing through a block of material known as a *gradient index material* (GRIN). This also gives an idea of why the blue light is scattered at sunset by the atmosphere to make the Sun appear as reddish. The blue light-wave is more energetic in its motion from side to

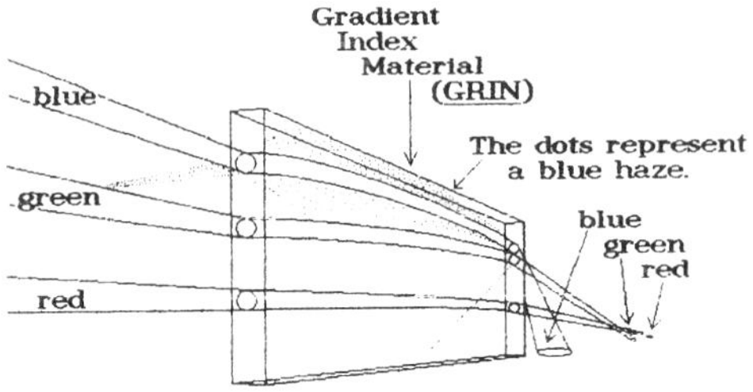

Figure 30

side. It is also interesting to note that the blue haze seems to collect back at one spot to exit the other side, as does the wavy light in Figure 12. The path of the blue light also seems to have a pronounced hump in the middle (see Figure 30).[3]

Notes

1. Center for Occupational Research and Development, (CORD), *Introduction to Lasers*.
2. Eugene Linden, "Can We Really Understand Matter," *Time* magazine, April 16, 1990.
3. Gary Stix, "Bug-Eyed Gradient-Index Lenses Bend Light to Order," *Scientific American*, November 1990.

Six

The Bomb

When the atomic bomb was first invented, it was called an implosion bomb because rather than explode it undergoes gravitational collapse, shrinking in on itself to the point of nothing, converting its entirety into energy. A nuclear implosion causes a characteristic blackout at ground zero at the time of ignition.

The concept behind an atomic implosion is based in the knowledge that when you leave quantities of heavy metals in proximity to themselves, heat is generated, accompanied by a transmutation of some portion of that element into another, heavier element that was not previously present.

The proportion of material undergoing this transformation and the amount of heat generated does not depend on the amount of time that it is left alone, but rather on the amount of material that is present. The greater the amount of material, the faster it will change. If you have too much of the refined uranium 235 confined to a small area, the heat generated will become unmanageable. This is what occurs in a nuclear meltdown. The core of the reactor overheats and melts a path for itself toward the center of the Earth.

The atomic bomb was designed to act on the principle of the theoretical black hole. Amazingly, the formation of this black hole does not drag the whole world into it. The builders of the first atomic bomb were not sure this would not happen.

One of the major engineering breakthroughs from these original bombs was the *Archie*. This was a small radar device for determining altitude. The bomb was designed to explode in the air rather than on impact with the ground. The elevated ignition creates greater devastation.

When watching footage of the first bomb tests, you first note the buildings smoldering from the heat energy. This is closely followed by the buildings being blown away from ground zero, then sucked back toward ground zero in the ensuing fire storm (see Figure 31).

The heat energy release by the bomb causes the intervening air to expand, causing the outward blast. The rising hot air then creates the vacuum that pulls everything back toward ground zero.

When viewing film on underground nuclear testing, it is startling to observe the ground falling away, leaving a huge depression in the earth. You expect to view the ground bulging outward, being pushed by tremendous pressure from the nuclear reaction. This reverse reaction is what physicists expected from a black hole. They did not know, however, if matter would stop being consumed in this collapse.

The first atomic bombs were too heavy to be economically

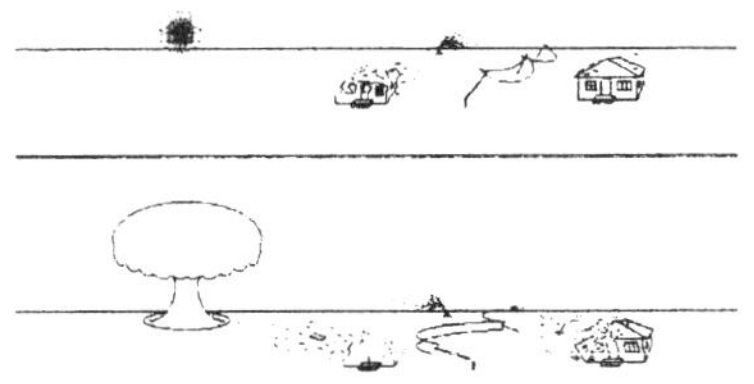

Figure 31

delivered via a long-range missile, but they were very simple in design. The Little Boy bomb was essentially a five-inch cannon that fired a uranium bullet with a velocity of 3,100 feet per second into three uranium target rings fitted to the muzzle (see Figure 32).[1]

A modern nuclear bomb is triggered by an implosion brought on by an accelerated particle being shot at a critical mass of uranium 235 or similar heavy unstable radioactive element. *Critical mass* means a mass in which a fission reaction has been initiated.

Free neutrons are essential in order to initiate a fission reaction. A slow neutron has to be introduced into a sufficient mass of uranium 235. The krypton and barium atoms resulting from this splitting of the nucleus are unstable, releasing more neutrons. Each resulting fission reaction produces an average of two and a half free neutrons. In 1934, Enrico Fermi, Emilio Segre, and their associates were the first to observe a slow neutron after having it pass through paraffin.[2]

The question is, why a slow neutron? Fast collisions cause

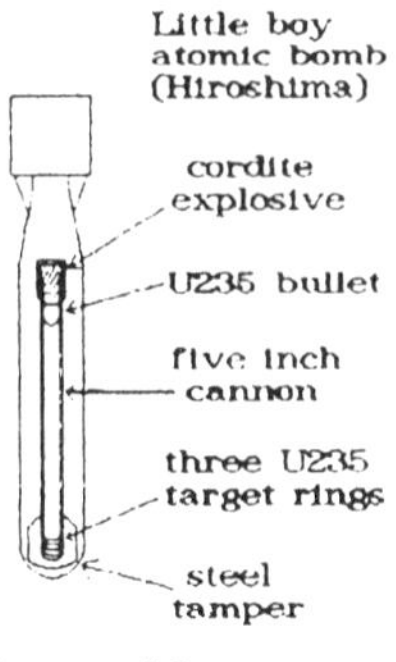

Figure 32

greater damage as a rule. If too much energy is used while extracting the neutron, it deteriorates into a proton and an electron. The proton, although massive enough to cause similar damage to a nucleus as the neutron, is restricted from doing so due to its electrical charge guiding the proton by the rules governing magnetic forces. This deflects a free-moving proton from hitting the next target nucleus by being magnetically repelled from protons and attracted toward the electrons, thus causing an energy loss in any future proton-nucleus collisions.

It has been experimentally shown that a neutron can be separated into a proton and an electron plus an immense amount of energy in high-impact particle colliders. A neutron removed from the nucleus has an average life-span of about seventeen minutes.

The neutron has slightly more mass than a proton, and is almost two thousand times more massive than an electron. From the reaction of a neutron in a collision with another neutron, it acts as though it possesses a great deal of angular momentum. Imagine a fast-rotating neutron moving in slowly and touching another spinning neutron—you would expect it to skid off, whereas, in more substantial encounters, the neutron would borrow energy from the angular momentum and gain velocity. The neutron does have a magnetic moment that suggests it is an entity of fast-moving electrical charges.

The first particle accelerators were developed by E. O. Lawrence from an idea he acquired by reading an article written by R. Wideroe. It consisted of two D-shaped chambers of an antimagnetic material, formed so the atomic particle projectile wouldn't make contact with its surface (copper and aluminum are both antimagnetic, repelling both positive and negative particles). The guides were called "Ds" due to their shape and were placed in the proximity of two magnets that kept the charged particles emitted naturally by a radioactive material moving in circular paths. A charged particle naturally travels in circles in the presence of a magnetic field.

An ion source near the center releases protons or deuterons (ions of deuterium). The chambers were subjected to a radio frequency that pulled, then repelled, the magnetic particles. The particles gained energy in steps each time they crossed the gap between the Ds. This, in turn, influenced the trajectory of the particles to be an outward spiral (see Figure 33).

For accelerating particles in a nuclear bomb, the cyclotron has been replaced by the maser wave guide, known as the *klystron triggering device*. The klystron works on the same principle as the cyclotron. It carries the charged particle out with the amplified radio signal. The wave guide is designed to accentuate the radio signal. The width represents a multiple of either a half or whole magnetic length. The height, as depicted, represents a quarter of the magnetic wave, which curiously is more effective at canceling a wave than having the standing wave pattern being 180 degrees out of phase. The bell-shaped lip at the end of the cavity aids the escape of the signal from the mouth. The klystron used in a nuclear bomb is tubular, but the height of the chamber is still at a multiple of the magnetic wave (see Figure 34).

The holes in the screen that allow you to look into a microwave oven are a multiple of one-quarter wave, of the energy-wave size, in order to stop harmful electromagnetic radiation from escaping.

There is much more to consider if you intend to build a bomb. The casing material even plays a role in promoting a successful fission reaction. In a reactor core, cadmium rods are introduced in order to slow the fission reaction. When the neutron bounces off from a cadmium atom, it gains energy in the form of velocity. Beryllium has been selected as the casing material because of its strength, heat resistance, and its ability to reflect the neutrons back inward without adding velocity. Beryllium also has an excess of neutrons, which it donates to the

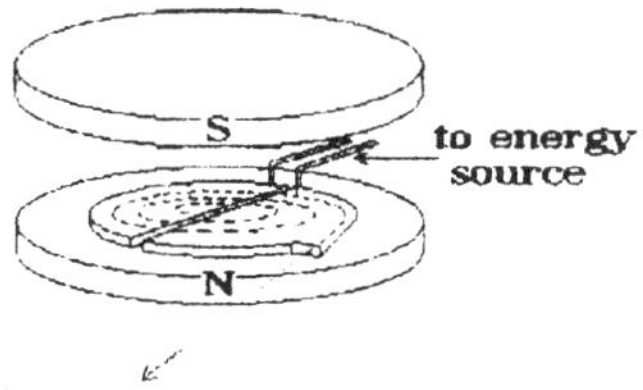

Figure 33

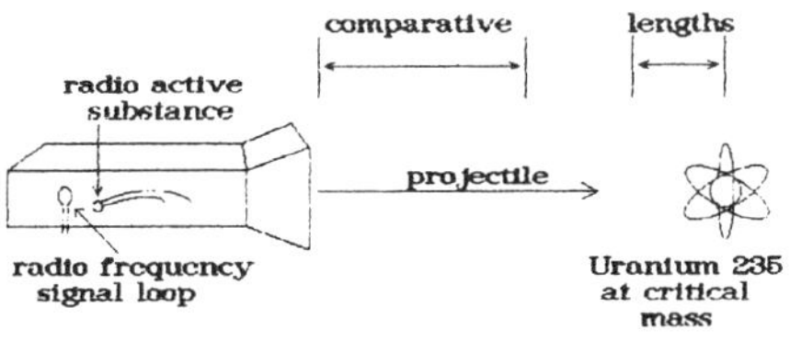

Figure 34

reaction. Designing a bomb is not the topic I want to discuss, however, I only want to convey the concept of matter collapsing into energy.

There is an instant during fission at which the krypton and barium, both being positively charged, are under tremendous electrostatic pressure. At this instant, they occupy the focal point of a tremendous amount of energy.

The charged-particle projectile from its faster time-rhythm at near its speed of light exceeds the heavier mass' speed of light; this causes the evolutionary process of the larger mass to accelerate, creating a small black hole. The hard part to comprehend is that the heavy mass' time stops at that instant, producing an extremely long, skinny photon because of the differences in time and pressure.

Imagine it as a dropped coin that lands on edge. As it progresses to a flat position, the edge bounces up in a rotary motion, going around the edge. The coin also turns laterally through its surface orientation, though not as fast as the edge is bouncing. The edge will bounce several times around before the surface makes a rotation. As the altitude of the bounce decreases, the gyrations become more rapid. This reminds me of the Schrodinger wave function. The decrease in the altitude of the bounce is equivalent to what the outside observer is witnessing as it relates to size. The lateral movement is directly proportional to the time scale for the inside observer. When the coin comes to rest, time freezes, and this is the instant the heavy mass produces a long, steady, unvibrating state that we perceive as gravitational collapse.

Between the proton projectile and the heavy mass there is no way of distinguishing which has exceeded the speed of light. Space may have a relationship to its adjoining segments similar to the right-hand rule of electricity: one segment in one direction, the next perpendicular to it, and the third perpendicular to both. When a proton of sufficient energy enters the vicinity, it could trigger long segments to reorient themselves into the longer-space version and raise havoc with the Pauli exclusion principle. Matter seems to need the electrons in order to regulate its dissipation of mass into energy. Electrons seem to be a byproduct of the process itself. This—compounded by the atom's loss of neutrons, which stabilize the atom while it's under tremendous pressure—causes its collapse.

Outside the Schwarzschild radius, matter is absorbing long-wave photons. This pulls matter inward but also creates a heat barrier, limiting the amount of outside matter pulled in. The long-wave photon also strips the outer-shell electrons from surrounding atoms, creating a situation where electronic devices will not work, not even direct current—which, for instance, would keep you from starting your car.

The ignition of hydrogen in a nuclear bomb is brought about by having the hydrogen contained in a canister designed as a microwave cavity near an atomic implosion. The X rays generated by the implosion distort the hydrogen's fragile perception of time, similar to the photon distortion described in the laser (see Figures 35 and 36).

The initial nuclear implosion is characterized by a total blackout at ground zero. The atoms send out their long waves, which appear as black.

The accompanying local area electrical disturbance also brings about a condition where having a second nuclear implosion would be impossible for a length of time. It is necessary to produce a radio-frequency signal in order to accelerate the proton inside a klystron, to initiate the reaction. The abundance of local energy would also sour the fission reaction.

This would make a clean neutron bomb an ideal defensive weapon due to its small kill radius of 1500 yards and short contamination period of thirty days. Before incoming ICBMs can hit the ground these could be detonated above their targets, nullifying them. This is probably why the Russians recently

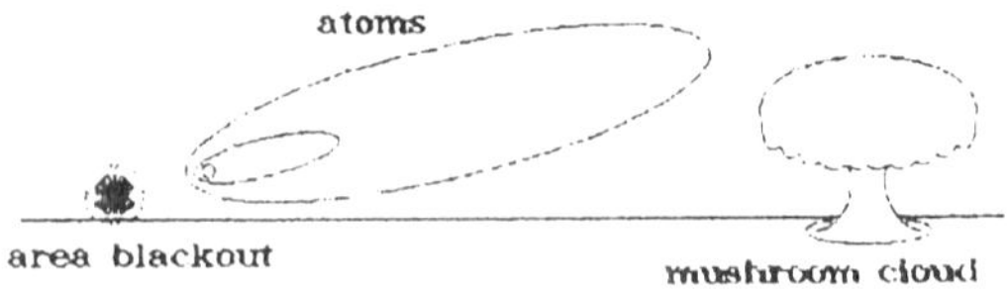

Figure 35

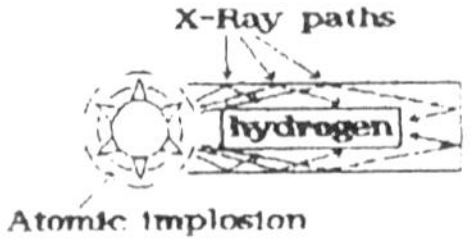

Figure 36

deployed antiballistic nuclear missiles around their cities, in violation of the Strategic Arms Limitation Treaty.

The klystron should not be confused with the krytron. The krytron was an antiquated optical triggering device used in the *Fat Man bomb* that was dropped on Nagasaki. The krytron precisely lit the cordite explosives, which compressed the plutonium core from all sides at the same instant of time, which is all that's needed to initiate a reaction from the naturally fissionable plutonium.

I believe that light waves and radio waves are related only to the extent that one is a natural reaction to the other. I think the time-rhythm conditions that cause you to see the black spots on the Sun represent the size and shape of atomic vibration that end up looking black, because of their extended length. The Sun's vibration condition (length-to-width ratio) creates a positive charge that can extend through the electrically conductive vacuum of space to cause a disruption in radio signals, even electrical storms, here on Earth.

Notes

1. Richard Roads, *Making of the Atomic Bomb* (New York: Simon & Schuster, 1986).
2. Emilio Segre, *From X-rays to Quarks* (1980).

Seven
Saturn

The space probe *Voyager* that went past Saturn discovered some very unusual events happening in the rings that circle that planet. One extraordinary example of these events is the spokes that stand straight out from Saturn through all of its rings, staying on the side away from the Sun. Another strange phenomenon that seems incomprehensible was the *F-Ring*. Saturn's rings resemble the grooves on a phonograph record and have letter designations. The F-Ring is actually braided in 150-mile segments without the elements of the ring making contact with each other. The ring appears to consist of water vapor and iron particles. The only hypothesis expounded so far is the shepherd-moon theory, which makes absolutely no sense. This theory states that the influences of two nearby moons' gravitational pull guides the strands of the ring in such a way as to braid them. Within conventional theories of gravity, the pull should completely blend this ice-crystal, iron-dust mist together. In the book *The Planets: A Journey into Space*, by Eric Flaum, I was reading that the F-Ring has up to ten distinct strands within its braids. In the pictures published in *The Detroit Free Press*, I could only see three (see Figure 37).

In order to understand these events, we should try to link them to something within our realm of experience. The braiding of the F-Ring seems such a mystery. The only links I can think of are the way electrons exit a hot oven, orienting themselves in three positions, and the three distinct positions in the flight of a Frisbee. The iron particles have acquired some property that

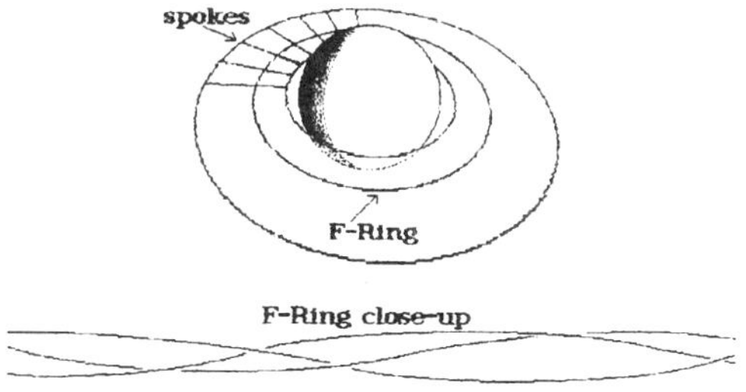

Figure 37

should be similar, but is somehow more than magnetism. These actions cannot be explained within the bounds of magnetism as they currently are understood. Some time ago, in a book written about Tesla's research, I ran across something that reminds me of the spokes striking out from the leeward side of Saturn. I will devote the next chapter to that information.

Eight
Tesla

This entire chapter is taken directly from the book *Inventions, Researches, and Writings of Nikola Tesla*, by Thomas Commerford Martin, starting at page 226.[1]

I think it best at this juncture to bring before you a phenomenon, observed by me some time ago, which to the purely scientific investigator may perhaps appear more interesting than any of the results which I have the privilege to present to you this evening.

It may be quite properly ranked among the brush phenomena—in fact, it is a brush, formed at, or near, a single terminal in high vacuum.

In bulbs provided with a conducting terminal, though it be of aluminum, the brush has but an ephemeral existence, and cannot, unfortunately, be indefinitely preserved in its most sensitive state, even in a bulb devoid of any conducting electrode. In studying the phenomenon, by all means a bulb having no leading-in wire should be used. I have found it best to use bulbs constructed as indicated in Figs. 141 and 142.

In Fig. 141 the bulb comprises an incandescent lamp globe *L*, in the neck of which is sealed a barometer tube *b*, the end of which is blown out to form a small sphere *s*. This sphere should be sealed as closely as possible in the centre of the large globe. Before sealing, a thin tube *t*, of aluminum sheet, may be slipped in the barometer tube, but it is not important to employ it.

The small hollow sphere *s* is filled with some conducting powder, and a wire *w* is cemented in the neck for the purpose of connecting the conducting powder with the generator.

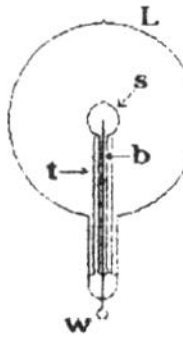

Fig. 141

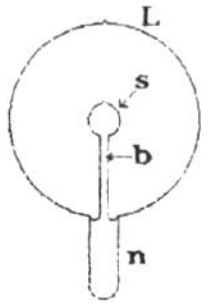

Fig. 142

The construction shown in Fig. 142 was chosen in order to remove from the brush any conducting body which might possibly affect it. The bulb consists in this case of a lamp globe *L*, which has a neck *n*, provided with a tube *b* and small sphere *s*, sealed to it, so that two entirely independent compartments are formed, as indicated in the drawing. When the bulb is in use, the neck *n* is provided with a tinfoil coating, which is connected to the generator and acts inductively upon the moderately rarefied and highly conducted gas inclosed in the neck. From there the current passes through the tube *b* into the small sphere *s*, to act by induction upon the gas contained in the globe *L*.

It is of advantage to make the tube *t* very thick, the hole through it very small, and to blow the sphere *s* very thin. It is of the greatest importance that the sphere *s* be placed in the centre of the globe *L*.

Figs. 143, 144 and 145 indicate different forms, or stages, of the brush. Fig. 143 shows the brush as it first appears in a bulb provided with a conducting terminal; but, as in such a bulb it very soon disappears—often after a few minutes—I will confine myself to the description of the phenomenon as seen in a bulb without conducting electrode. It is observed under the following conditions:

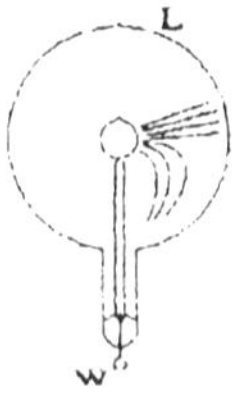

Fig. 143

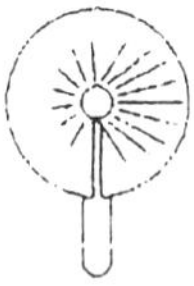

Fig. 144

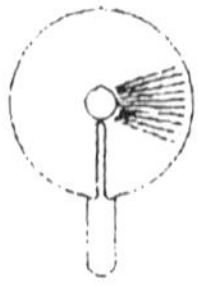

Fig. 145

When the globe *L* (Figs. 141 and 142) is exhausted to a very high degree, generally the bulb is not excited upon connecting the wire *w* (Fig. 141) or the tinfoil coating of the bulb (Fig. 142) to the terminal of the induction coil. To excite it, it is usually sufficient to grasp the globe *L* with the hand. An intense phosphorescence then spreads at first over the globe, but soon gives place to a white, misty light. Shortly afterward, one may notice that the luminosity is unevenly distributed in the globe, and after passing the current for some time the bulb appears as in Fig 144. From this stage the phenomenon will gradually pass to that indicated in Fig. 145, after some minutes, hours, days or weeks, according as the bulb is worked. Warming the bulb or increasing the potential hastens the transit.

When the brush assumes the form indicated in Fig. 145, it may be brought to a state of extreme sensitiveness to electrostatic and magnetic influence. The bulb hanging straight down from a wire, and all objects being remote from it, the approach of the observer at a few paces from the bulb will cause the brush to fly to the opposite side, and if he walks around the bulb it will always keep on the opposite side. It may begin to spin around the terminal long before it reaches that sensitive stage. When it begins to turn around, principally, but also before, it is affected by a magnet, and at a certain stage it is susceptible to magnetic influence to an astonishing degree. A small permanent magnet, with its poles at a distance of no more than two centimetres, will affect it visibly at a distance of two metres, slowing down or accelerating the rotation according to how it is held relatively to the brush. I think I have observed that at the stage when it is most sensitive to magnetic, it is not most sensitive to electrostatic, influence. My explanation is that the electrostatic attraction between the brush and the glass of the bulb, which retards the rotation, grows much quicker than the magnetic influence when the intensity of the stream is increased.

When the bulb hangs with the globe *L* down, the rotation is always clockwise. In the southern hemisphere it would occur in the opposite direction, and on the equator the brush should not turn at all. The rotation may be reversed by a magnet kept at some distance. The brush rotates best, seemingly, when it is at right angles to the lines of force of the earth. It very likely rotates, when at its maximum speed, in synchronism with the alternations, say, 10,000 times a second. The rotation can be slowed down or accelerated by the approach or receding of the observer, or any conducting body, but it cannot be reversed by putting the bulb in any position. When it is in the state of the highest sensitiveness and the potential or frequency be varied, the sensitiveness is rapidly diminished. Changing either of these but little will generally stop the rotation. The sensitiveness is likewise affected by the variations of temperature. To attain great sensitiveness it is necessary to have the small sphere *s* in the centre of the globe *L*, as otherwise the electrostatic action of the glass of the globe will tend to stop the rotation. The sphere *s* should be small and of uniform thickness; any dissymmetry, of course, has the effect to diminish the sensitiveness.

The fact that the brush rotates in a definite direction in a permanent magnetic field seems to show that in alternating currents of very high frequency the positive and negative impulses are not equal, but that one always preponderates over the other.

Of course, this rotation in one direction may be due to the action of the two elements of the same current upon each other, or to the action of the field produced by one of the elements upon the other, as in a series motor, without necessarily one impulse being stronger than the other. The fact that the brush turns, as far as I could observe, in any position, would speak for this view. In such case it would turn at any point of the earth's surface. But, on the other hand, it is then hard to explain why a permanent magnet should reverse the rotation, and one must assume the preponderance of impulses of one kind.

As to the cause of the formation of the brush or stream, I think it is due to the electrostatic action of the globe and the dissymmetry of the parts. If the small bulb *s* and the globe *L* were perfect concentric spheres, and the glass throughout of the same thickness and quality, I think the brush would not form, as the tendency to pass would be equal on all sides. That the formation of the stream is due to an irregularity is apparent from the fact that it has the tendency to remain in one position, and rotation occurs most generally only when it is brought out of this position by electrostatic or magnetic influence. When in an extremely sensitive state it rests in one position, most curious experiments may be performed with it. For instance, the experimenter may, by selecting a proper position, approach the hand at a certain considerable distance to the bulb, and he may cause the brush to pass off by merely stiffening the muscles of the arm. When it begins to rotate slowly, and the hands are held at a proper distance, it is impossible to make even the slightest motion without producing a visible effect upon the brush. A metal plate connected to the other terminal of the coil affects it at a great distance, slowing down the rotation often to one turn a second.

I am firmly convinced that such a brush, when we learn how to produce it properly, will prove a valuable aid in the investigation of the nature of the forces acting in an electrostatic or magnetic field. If there is any motion which is measurable going on in the space, such a brush ought to reveal it. It is, so to speak, a beam of light, frictionless, devoid of inertia.

I think that it may find practical applications in telegraphy. With such a brush it would be possible to send dispatches across the Atlantic, for instance, with any speed, since its sensitiveness may be so great that the slightest changes will affect it. If it were possible to make the stream more intense and very narrow, its deflections could be easily photographed.

Rather than shepherd moons being responsible for the antics of the F-Ring (as discussed in the previous chapter), I would suspect the shyness of the spokes indicates an affinity to Tesla's brush—this in turn charges the iron particles, and possibly, given the remoteness and vacuum conditions, emulates an actual giant electron standing wave pattern.

Note

1. Thomas Commerford Martin, *The Inventions, Researches and Writings of Nikola Tesla* (1894; reprint, Milwaukee, WI: Lee Engineering Company, 1952).

Nine
Electrons

Albert Einstein and Leopold Infeld write in their book *The Evolution of Physics* a very eloquent description of the chance discovery of electricity by Galvani, followed by Volta's Voltaic cell battery. Their treatment of the subject is inspiring. I advise everybody to read it in conjunction with this book. The development of the concept of electrical field as narrated by them is indispensable. I'm going to skip through an outline of their ideas, hoping the reader will follow this advice.

Hans Christian Orsted discovered that when electric current flows through a wire it affects the orientation of a compass placed near the wire. The electric current does not direct the needle toward the wire; instead, the needle points at right angles to the wire. This tied electricity and magnetism together for the first time. They were formerly considered separate phenomena. This is also the first force field perpendicular to its source. Newton's gravity and Coulomb's magnetism act along the line joining the two attracting or repelling bodies. Experiments by Rowland discovered that a moving magnet induces a current into nearby wire and the potential difference is directly proportional to the velocity of the magnet. The potential difference being tied to the relative velocity of the magnet respective to the wire is important because the energy of the velocity is directly translated into voltage. The whole mechanical point of view was based on the belief that all phenomena can be explained in terms of forces, depending only on the distance and not on the velocity (see Figure 38).

Figure 38 represents a wire hoop that is conducting an electric current, which is passing through a paper cutout sprinkled with iron filings. The iron filings form circles corresponding to where the magnetic field is strongest. The arrows have been added to show the direction of the polarity. Note that the arrows drawn to show the direction of the current point from the positive toward the negative. This method of denoting polarity is a holdover from the original electrical theories, in which the founders arbitrarily guessed at the direction of the current.

The first analogy to help visualize electricity was to compare electricity to water. In this comparison, the voltage was equivalent to the pressure and the current in amperes is the volume passing a given point per unit of time.

Schrodinger's wave function adds to the mystery surrounding electrical theory. This actually starts with Louis de Broglie, who postulated that an electron acted as if it were a wave. Picture

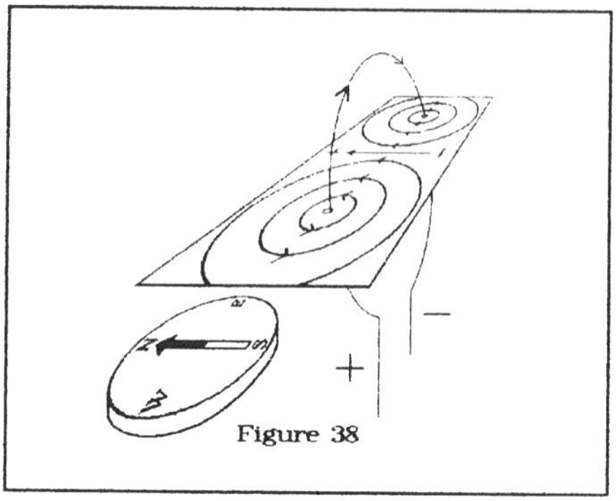

Figure 38

a transverse-wave as holding one end of a taut rope that is tied to a pole, then give the rope a jerk. A vibration wave will travel down the rope to the other end and if it's tight enough, you will see it traverse the line several times. The other type of wave that we can visualize is the longitudinal wave. The longitudinal wave is like a sound wave traveling through air—a thick, compressed wave front followed by a rarified segment of air. There is one important parameter for the Schrodinger wave. Its existence is limited to even numbers of waves; otherwise it cancels itself out.

De Broglie discovered another astonishing connection between the wave and the particle for each of Bohr's orbits. The electron's momentum equals Planck's constant divided by the standing wave's wavelength. Each electron orbit is a standing wave pattern.

While this was being speculated about in Europe, Clinton Davisson of Bell Laboratories in the United States noticed peculiar wave-interference patterns caused from electrons reflected by nickel crystals. These experiments verified the de Broglie wave, and interference patterns were soon after observed in neutrons. This observation was followed by the discovery that all matter was capable of producing wave patterns. The conclusion that can be drawn from matter-producing wave patterns is that all matter is in a state of vibration.

The Schrodinger wave function is a mathematical description of the electron's wave. There is only one wave for all the electrons of an atom. This seems to create problems for the Bohr orbiting-electron theory, due to all the varying electron-shell orbit levels; at the atom's pulse rhythm, all electrons have to move to their next position simultaneously. Within this Schrodinger theoretical wave-framework, the electron expands exponentially the longer it is removed from an atom, so that, within a few seconds after being removed, the electron would be larger than a football field.

To this day, the predominant theory of electricity is that

there is an abundance of electrons gathered around one pole of a battery, waiting for a conductor to carry them to the other pole, which has a shortage of electrons. The obvious problem with this is the question of why electrons are limited to this transfer within the one battery, instead of jumping to another battery, when only one set of opposite poles is connected. When the batteries are linked together in a positive-to-negative manner, this pattern is known as a *series connection* and doubles the voltage across the remaining terminals.

In order to give a believable alternative to current theories, I have to work within this baffling framework, all of which I still haven't listed. Rutherford established the planetary-atom model by bombarding gold foil with protons and tracing the proton scattering. This experiment gave the following results: The positive charge and 99% of the mass is located in an area the shape of a ball at the center, occupying only one ten-thousandth of the diameter of the atom. It seems theoretically impossible to compress the positive charges of all the protons into this small area. The force that was once thought to keep the electrons away—to keep them from being dragged into the nucleus—was centripetal force. Maxwell's calculations, predict that an electron performing such a motion should be radiating its energy and should crash into the nucleus in less than a second. Neils Bohr put a Band-Aid on this problem by decreeing that there are allowed orbits that an electron could occupy without radiating its energy. There is a certain energy associated with these orbits, and an electron will give up energy only when it drops from an outer orbit to an inner one. The energy given up will equal the difference of energy between these orbits. His observations were accurate, but I feel his theory does not address the problem. The scientific community has long since thrown out the Bohr orbiting-electron model, but there has not been another acceptable theory put forward to replace it yet.

Looking at the path of an electron in a bubble chamber gives some clues as to what the make-up of an electron could be. Figure 39 shows a proton entering a bubble chamber and displacing particles. The path of an electron that has been knocked out of an atom resembles the path of a spinning jack. If contact is made with the center of momentum, the jack should take off in almost a straight line and decay rapidly. Instead, its rotation decays in leisurely spirals. This seems to indicate contact with the electron is limited to a small point being driven by a larger centripetal momentum. Although there seems to be ample evidence for an electron being a particle, keep an open mind—an electron may be a localized effect caused by surrounding disturbances (see Figure 39).

There are four basic electronic component categories that, when put together in different combinations, make up all our present electronic equipment. The category titles are: resistor, coil, capacitor, and rectifier. The function of each category is as follows:

Resistors restrict the current, measured in amps, through a circuit. The resistor can be used to step down the voltage in a circuit by tapping in between two series-connected resistors. Each resistor drops a certain voltage with respect to its value,

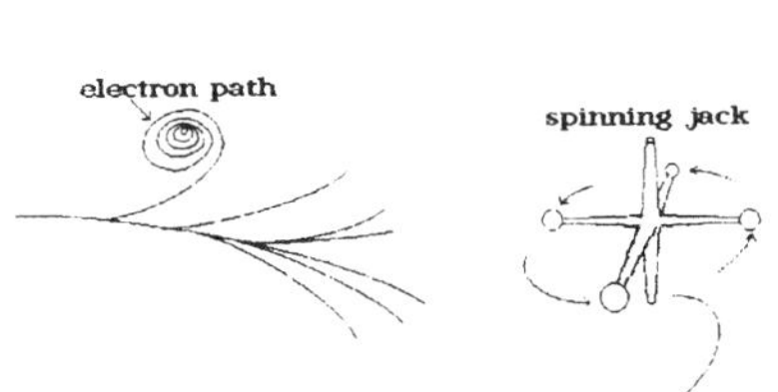

Figure 39

measured in ohms. The voltage will be divided in a ratio proportional to the resistance ohm's value.

A *coil* is simply wire wound into a coil. Direct current passes through readily. When alternating current is applied to a coil, the expanding and contracting magnetic field requires some time to change polarity. This time element causes resistance to alternating current, in ohms, that increases with respect to increasing frequency.

The *capacitor* is constructed by having two conducting plates facing each other across an insulating barrier. As one plate becomes charged, the other develops an opposite charge in response to the initial charge. Because no contact is made between plates, direct current is blocked while alternating current passes freely. As the frequency increases, the capacitor becomes a more efficient conductor and will reach a stage where it has almost no resistance to the alternating current.

The *rectifier* is simply layers of material that responds to one polarity only for each layer. When alternating current is applied, one layer becomes charged with one polarity, causing the other to become oppositely charged. A rectifier, as its name implies, allows passage of current in one direction only, filtering out most reverse-current flow.

The mode of operation of the capacitor belies the particle theory of electrons passing through a wire. If electrons bunching on one plate of a capacitor repelled electrons from the other, the associated actions of electrons within a battery should be capable of moving from a negative battery pole of one battery to the positive pole of another without connecting the remaining poles of the two batteries.

Rutherford's first notable experiments as a teenager in New Zealand established that electrical current runs on the outside of the wire. Rutherford did this by magnetizing an iron plate, then, using acid, removed the outer layer of iron. The magnetism was removed with the outer layer. In this manner, Rutherford showed that magnetization was only a surface effect. Electromagnets that are composed of layered iron plates can attain greater magnetic fields because the surface area guides the field.

Circular magnetic flux lines form around every wire that is conducting electricity. Flux circles, or nodes, space themselves evenly down the length of a conducting wire. Electrons, or potential difference, might be nothing more than the nodes that lie on the wire as Gauss (flux) lines. Experiments performed by Maxwell show electromagnetic wave formations assume interlocking positions (as depicted in Figure 40). I do not believe that an electron enters a wire on one end, causing a chain-displacement reaction that results in an electron being displaced from the other end of the wire. The length of cable used between computer equipment is critical. The signal sent by one component can only be picked up by another component at a location on which a node forms. The lack of a detectable signal between nodes spawned the term *ethernet.* If the electron were really a charged particle you would expect a magnetic-mount antenna to stifle a radio signal's output (see Figure 40).

When two coils are in close proximity, in a one-to-two winding ratio, the circular flux lines are directed into larger circles surrounding the secondary coil. The flux lines station themselves, evenly spaced, along the wire and could affect the electric charge by the fact that they are stationary in spatial orientation, but possess a size that conflicts with the surrounding local time-scale (see Figure 41).

In Figure 41 the voltage is increased through induction in the secondary coil. An alternating current is applied to the primary coil. The expanding magnetic flux lines of the primary coil cut across the secondary coil and cause the current to flow in the secondary coil. This secondary coil has twice as many windings as the primary coil. It appears to the wire as having a denser node population, causing redistribution of nodes down the wire. The voltage can also be stepped down by reversing this

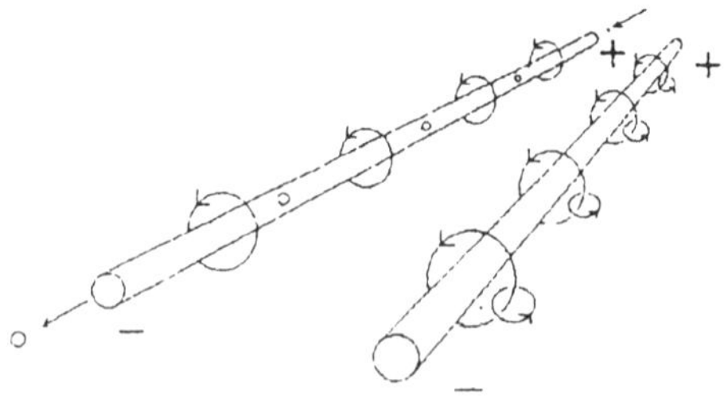

Figure 40

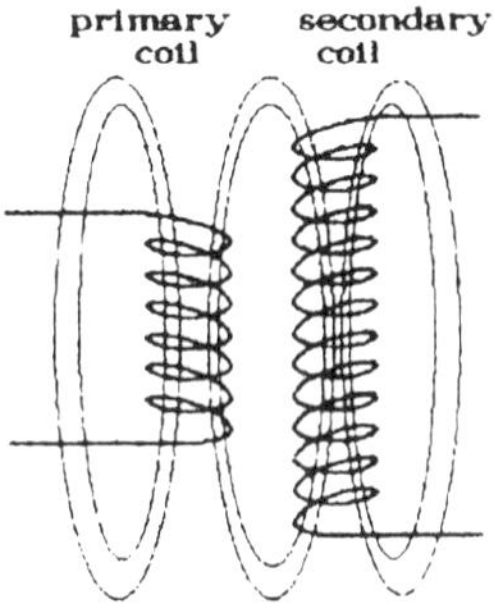

Figure 41

process—simply provide twice the windings in the input coil and you halve the voltage coming from the output coil.

The death knell for the particle theory of electricity should be the fact that decreasing the diameter of a wire by half would have to speed the volume of anything flowing through by four times in order to maintain the same amperage. The speed of electrons would therefore have to increase through the smaller wire in order to maintain the same current. As Rowland demonstrated, this would increase the magnetic field strength. This does not happen. When an equal current flows through a smaller wire, the magnetic field strength remains the same. If the electrons passed through a wire, a transfer of energy would take place between coils using direct current; you would expect that the magnetic field would travel down the wire in unison with the electron. Instead, you find the Gauss lines evenly space themselves along a wire and stay stationary.

My thoughts on electrons are that electrons consist of a swirl of ether that can capture photons within its circle, and keep them by being reflective to outside ether, in much the same way that fiber-optics cables carry light through twists and turns.

Every piece of matter is in a state of vibration. The vibration qualities control the electrical potential through the nature of their motion. *Positive* is represented by an atom's expanding-contracting vibration being greater than its side-to-side motion. Conversely, a greater side-to-side vibration would represent the *negative* charge. The nature of the vibration would also affect the time-rhythm of the atom, the positive charge being associated with the slower time-rhythm, and negative with the faster rhythm. The basis of these thoughts have their roots in my explanation of the positive charge associated with a nuclear implosion, and the positively charged solar wind. As I previously speculated in chapter 4, this could also allow for more electrons in the valence shell in faster time-rhythm locations.

Metal detectors work on a principle of sending out and

receiving reflected radio frequencies that can be tuned to specific signals so that you can discriminate against unwanted metals. It seems possible that any body that establishes itself as a unit with its shared time-rhythm creates a similar aura proportional to its size, with the frequency dependent on the composition. This may be further subdivided into smaller portions within the units, but the mass is equivalent to all that is enveloped by the shared time-rhythm aura.

The core of thermodynamics, and all energy redistribution, is to be found in the notes of a lecture given by Joseph Black around two hundred years ago.

> By the use of this instrument [thermometer] we have learned, that if we take one thousand, or more, different kinds of matter, such as metals, stones, salts, woods, feathers, wool, water, and a variety of other fluids, although they be all at first of different heats [temperatures], let them be placed together in the same room without a fire, and into which the Sun does not shine, the heat will be communicated from the hotter of these bodies to the colder, during some hours perhaps, or the course of a day, at the end of which time, if we apply a thermometer to them all in succession, it will point precisely to the same degree.

The gist of this argument is that an entire body will resonate at the same frequency. For example, if you have a truck loaded with sand piled on a flat surface, and the truck plus sand weighs two tons in its entirety, any object that is struck by that truck will feel the energy of two tons proportional to the truck's speed at the time of impact. The surface of the sand would have a different temperature due to its direct contact with an outside rhythm. You also get the same velocity reading when you focus radar on any portion of the truck or load.

The Earth's atmosphere is enveloped in an electrically charged coating called the *Van Allen belt*. The Van Allen belt in my theory is the result of the Earth's vibration state. This layer allows the passage of select light energies and filters out several bands of electromagnetic waves.

Knowing that there is a point beyond which satellites experience centripetal force leads to speculation that the Van Allen belt is probably the demarcation line beyond which an object experiences centripetal force. The Van Allen belt may act as more than a buffer for unwanted radio signals (see Figure 42).

It is easy to see the logic in the current explanation of the transfer of power with the accompanying increase in voltage across a coil. It is a lot harder to picture the voltage increase in the cascade voltage multiplier using conventional arguments. Each consecutive capacitor has to repel an electron to a higher energy on the opposing side; within the framework of particle theory, the primary electrons that remain isolated propel the secondary electrons to double their speed. The direction is restricted to one way by the diodes (see Figure 43).

An electron is linked in some manner to light, so that the wave segment, when accepted, raises the electron's potential. The propagation of an electromagnetic wave has been studied by Maxwell and found to consist of circular Gauss lines that always occupy planes perpendicular to their adjoining counterparts in such a manner that the wave's passage is three dimensional, still lacking a time element as compared to the two-dimensional passage of light.

Alternating current is easier to transmit than direct current. I use the word *transmit* because it seems possible to me that there is no difference between an electron and a magnetic field. The length of time for the rotation of any electron would also be contingent on the rotation-size of adjacent electrons in adjoining atoms sharing their time-rhythm.

An interesting point I read some time ago is that when you split a pair of electrons up and reverse the spin on one, as with a Stern Gerlach device, the other electron's spin reverses instanta-

Figure 42

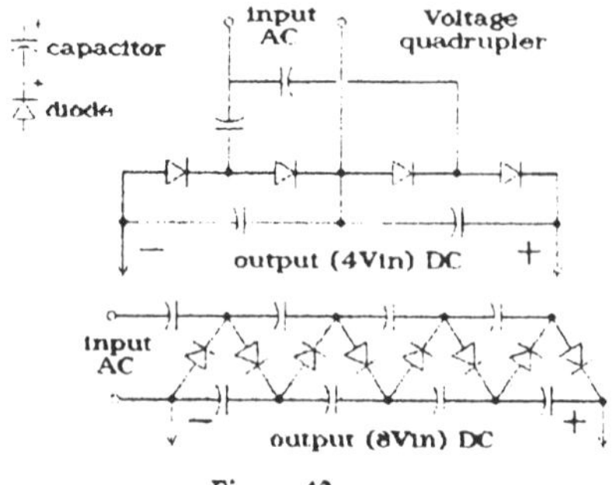

Figure 43

neously, faster than light. This brought about speculation on tachyon theories. Matter that has been intentionally stripped of electrons may develop a length-scale that would exceed the length-scale of matter that has not been subjected to the dropping of pressure and addition of heat, which is the standard method for removing electrons in cathode ray tubes. With the new length-scale, the electron's spin reversal would not really exceed the speed of light; it would only overlay the length-scale of intervening matter. Photons that are shorter in height find matter translucent. The larger electrons might undergo a change in form. Possibly as an electron extends its length, the height of its swirl diminishes. This would allow the electron to pass freely through intervening matter.

The two wires comprising a thermocouple are composed of different elements. Both wires are subjected to the same frequency of light. The two elements have differing reactions to the same frequency light, due to their differing mass, resulting in each wire landing in a different time scale, or vibration rate. The differing time scales would translate to electron spin, being greater horizontal to the surface, thus creating a negative difference in potential.

The round ball at the top of most flagpoles is placed there to protect the pole in a lightning storm. Even though it is taller than most of its surroundings, the ball shape acts as a capacitor and hinders the fast change of electrical potential. It would seem as though in order to change the potential of a capacitor, you have to change all of the electrons over its entire surface. The rate of change depends on the potential difference and is proportional to the surface area. Compare this to the lightning rod with its sharp point, which promotes the change of potential. The movement of an object horizontal to a single point is exaggerated, as compared to the same movement, when compared as a ratio, to a broad surface. When one section changes, the rest of the matter changes time-rhythms with it.

Electrons stripped from their atoms and shot from a hot oven through a small aperture have been found to exit in a regular order; the second is perpendicular to the first and the third is perpendicular to both of the preceding two electrons, reminiscent of the three positions in the flight of the Frisbee. What could cause the three possible positions of an electron passing from a hot oven? As I speculated with regard to the light passing through the hole in the box camera, the aperture, stimulated by light-waves, is only capable of four motions, back and forth, expanding and shrinking. Of these, only three count, because as the aperture moves forth in sequence with the electron, the reciprocal effect of this motion on the electrons is nullified.

When there is insufficient matter to confine the space-scale of an electron, the electron expands. This would cause a raised negative energy level in its faster time-rhythm. A case in point would be the energy gained when the electron crosses the gap in a cyclotron. The passage from the slower rhythm of the inner circles to the comparable faster time-rhythm of the gap causes the noticeable change in its charge. As the spark-gap increase—as in the Jacob's ladder experiment—the voltage increases also in small increments. It seems that by altering the size of an electron we affect the atom's time-rhythm.

The reaction to shortwave light creating negative charge in a photocell could be due to ultraviolet light's greater lateral motion from its greater width causing a greater pumping action horizontal to the surface. An electron will make a quantum jump any time a photon of sufficient energy presents itself. The electron quickly reestablishes its natural location with respect to the prevailing pressure, if allowed to do so; this briefly alters the atomic size and valence of an atom. As an atom emits a photon, it shrinks in size and returns to ground state (the electron valence charge that is normal for it). The surrounding atoms experience a vibration as they quantum jump to fill the vacancy left by this shrinking atom, they also develop new time-rhythms. The electron's return can be hindered by a thin layer of dielectric material (commonly copper nitrate). After jumping across this material, the electron cannot jump back, creating a potential difference across this material.

The groundwork I laid during my discourse on nuclear implosions has established my position that a positive charge is associated with the presence of long, narrow photons developed in a slower time-rhythm. Friction can be used to slow time-rhythms between two objects; this elongates and narrows the charge's shape. This longer swirl then requires more perpendicular room to travel. The difference of time it takes for the positive charge to dissipate can be used to collect enormous potential differences. The Van de Graaf generator is really a takeoff of the ice-pail experiment. A rubber belt continually adds the positive charge (extremely high potential differences can be reached—it has been used to accelerate protons). I believe that we have more than one time scale in opposite charges (see Figure 44).

For experimental purposes, Tesla made a Van de Graaf generator on a scale large enough to have a platform inside for a man to walk on. These generators were capable of producing several million volts, and Tesla, not one to leave anything un-

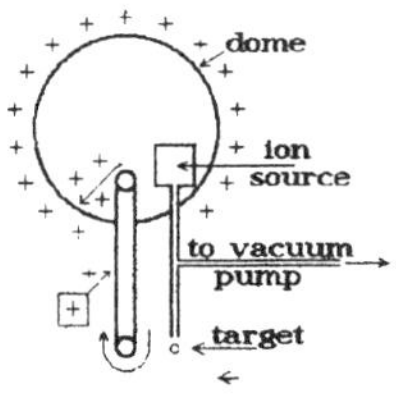

Figure 44

done, naturally sat inside one during operation. He relates that once inside there is no sensation, you can even touch the sides with no feeling of discomfort. Evidently the person inside adjusts to the slower time-rhythm.

The results of this experiment reinforce my inclination to discard current theories of electricity as a particle. The man inside the dome is accumulating the same charge as the dome and yet the electrical charges between the atoms within his body function normally. If electricity were due to a transfer of a particle, then touching the side of a conducting material whose atoms have several million volts potential should alter the valence of the matter coming in contact with it. The act of coming in contact with a huge electrical potential difference should alter the contacting material's molecular makeup. If charged particles are responsible for producing radio waves, a magnetic mount antenna should hinder radio transmissions. It would appear that the transfer of electricity, or radio signal, is indeed a shared time-rhythm phenomenon rather than an actual particle transfer.

Try thinking of an electron as a vortex of ether, which can be distorted by a passing photon. A factor in determining the polarity of the charge is the direction in which the electrons are distorted. When direct current is flowing through a wire, the magnetic-flux lines encircling the wire become gigantic. The electrons stretched horizontally to the surface could represent the negative charge. When the electron is stretched vertically from the atom it appears to be from a slower time-rhythm. This would make the positive charge the dominant polarity because, as in the Michelson-Morley experiment, it would take a slightly longer period of time to make the transition in the direction of motion (see Figure 45).

Two simple electronic devices, the capacitor and coil, demonstrate differing reactions to a sinusoidal voltage source. The capacitor is constructed of two conducting plates that face each other. The coil is simply wire wound into a coil. Two coils in close proximity to each other create a transformer. A transformer is sometimes used to alter the voltage in a circuit. The ratio of the number of turns in the primary winding, compared to the turns in the secondary winding of a transformer, determine the amount of voltage produced in the secondary winding. The capacitor has no contact between the facing plates; therefore, the capacitor blocks any direct current while it acts as a high-frequency short. The capacitor acquires a charge on opposite plate slowly, as compared to a transformer, which transfers a charge from the primary coil to the secondary coil as fast as the expanding magnetic flux lines cut through the secondary coil. A single coil passes direct current with little resistance, but resists high-frequency alternating current. When the reactance of a coil versus a capacitor is represented graphically, the direction of polarity is opposite and nullifies the reactance of the other.

The probability is that one conducts the slow-moving positive charge and the other the fast-moving negative charge. The ball acting as a capacitor resisting fast change of potential on a flagpole points toward the capacitor transferring the positive. Now consider the molecules of a dielectric material. The molecules are probably cone-shaped and all facing in the same direction. The flat side acts as a capacitor, which gradually builds a charge, resisting a fast change of potential, while the point-shaped end promotes the transfer of a negative charge.

The photon distortion from matter in a gravitational field would resemble that in the third example of Figure 45 and would be bunched up on the side facing the gravity source. Similarly, the trailing edge of a satellite would occupy a slower time-rhythm. Tesla noted that one charge preponderates over the other; I feel this would also apply to time. In my studies with light, I have found it expedient to treat its passage as a line that zig-zags back and forth along its path. Suppose light somehow became stuck making predominantly left turns when it came in

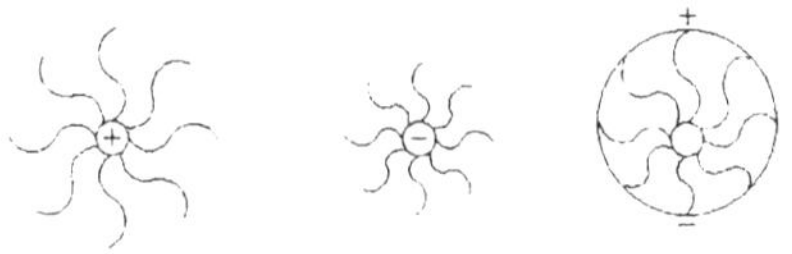

Figure 45

close proximity to an atom. Possibly layers of ether possess different refractive indices at different levels.

Draw a circle and then draw a square surrounding the circle, with contact between the two figures at four points, then draw another square within the circle, a square whose corners also touch the circumference of the circle. The ratio of one side of the inscribed square divided by the length from one side of the circumscribed square is equal to the square root of two. If we draw a diagonal line through the inner square, the hypotenuse of the resulting triangle is the length of one side of the outer square. It is amazingly coincidental that these two irrational numbers, pi and the square root of two, have this much in common (see Figure 46).[1]

An electron is assumed to be revolving because of its angular momentum, which is a product of pi. The two energies that we encounter when dealing with an electron are also a derivative of pi for its angular momentum, and the square root of two when we measure its shell energy. Possibly the electron voltage represents the measure of warp space experiences when a straight traveling photon encounters a circular photon.

It seems possible that a magnetic field is composed of rotary

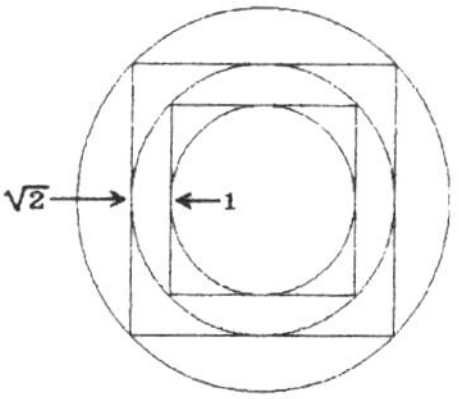

Figure 46

photons. Photons, having been emitted, radiate in all directions and are only lopsided in relation to the side toward the greater mass—and then it's only semantics. By rotary, I mean it goes in one direction and then closes its own circuit in an actual whirlpool. This electron model can absorb a photon, lengthening its rotational size, shifting it to a faster time-rhythm (from the faster time-rhythm perspective, it moves farther per pulse). The passage of an atom through space would stimulate an electron's rotary motion, which would also translate into a faster time-rhythm for that matter. Higher frequency blue light would be more energetic side to side, causing lateral distortion and creating larger electrons. This magnetic wave would still retain its individuality from light because light radiates in a different manner, as described earlier. The method of transmission of charge for a magnetic wave would be the enlarging one of the two electron dimensions. Imagine, if you will, that the electrons are more or less stationary in relation to the surrounding atoms, and that the nucleus is bobbing and weaving around, passing photons acting as a bellows and pumping energy into the electrons (like the two parallel power lines allowing the passage of light). It would seem to me just as convenient to move the center of an atom as it would the edges, especially when the edges are made up of energy and the quantum jumping is due to the passage of same. The examples of matter organizing in patterns under laser light reinforce this hypothesis.

The interaction between light and matter appears more complicated than I earlier stated. We have my earlier picture of matter bounding out of the path of light. At the atomic level, I would expect the absorption of light to behave as a simple on-or-off choice for the target matter. Bohr stated that light would be absorbed if the wave length was acceptable. I want to digress and revisit the concept of color. The combination of red, green, and blue pigments rejects all light to appear as white. Yellow, cyan, and magenta together absorb light to appear as black. The value for the sum of the waves is equal, but the difference is like adding positive numbers compared to adding negative numbers. This happens even though you can derive the additive primary colors from the subtractive primary colors and vice versa. It takes three factors to decide whether the matter is on or off. Because matter demonstrates three possible positions in the flight of the Frisbee, I feel there is an underlying physical law incorporated in the combination of three colors.

An electrical charge is determined by the size and plane of distortion of the electron swirl. The ether would be contained within its swirl because to venture outside would be to enter a new medium at too great an angle. Light penetration is limited between mediums, as far as angle is concerned. The transmission of an electrical charge happens in a rotary fashion. Lightning, however, seems to be a cross between both methods of travel (light and electrical) because of its apparent jagged course.

Truck drivers talking on CB radios often use a method of amplifying their signal to prioritize their conversation over everyone else's. This method is known as using a *linear* signal. The linear wave is strongest toward the front and rear of the truck sending the transmission and relatively weak to the right and left sides of the vehicle, which generally gives it greater reach down the road the truck is traveling.

The method of utilizing a linear signal is simple: Just install one antennae on each of the rear-view mirrors. The signal is the same frequency in all directions. There is one characteristic to determine which direction would possess the stronger signal; the two antennae determine which direction is controlled by the dominant, positive portion of the electromagnetic wave. The wave projected perpendicular to the front of the truck would be as if it were from the flat capacitor plate surface.

Another unexpected use utilizing moving magnetic fields is to be found in separating nonferrous metals. This employs a rapidly rotating ring of magnets in close proximity to a chute with nonferrous metals sliding down it. The differing amount of electrical resistance found in these assorted materials results in different trajectories for each type of metal, thus functioning as an automatic sorting device. If you have ever stood in the presence of a powerful transformer you would have felt as if everything was vibrating without having anything actually moving. This is the same principle and likewise causes a humming sound coming from the alternating field.

Note

1. Pete Jorgensen, "The Circumference to Diameter Ratio through the Hippocratean Relationship of Triangles to the Circle," unpublished research paper.

Ten

The Beginning of the Universe

Using your imagination, picture deep space. This is where matter has a very sparse population—only three atoms per cubic centimeter, as compared to the number present here on Earth, which contains some thirty-billion billion atoms per cubic centimeter. Matter composed mostly of hydrogen waits millions of years for the arrival of a spiral arm of a galaxy where stars are known to form. When the galactic arm finally arrives, it jostles the atoms together. This compresses the atoms into huge gas clouds that further condense into primeval stars. The Pleiades is an example of this kind of formation. These are the conventional ideas of star birth.

If the solar system condensed from a gas cloud into the present star and planet system that we are familiar with today, you would expect the gas condensation theory to allow the material composing the planets to blend into homogenous spheres. If somehow we managed to overcome these obstacles and still have isolated deposits of elements after condensing them from vapor, the planets should have a few aberrations insofar as shape is concerned. One of the moons of Mars is oblong, as if it were a captured chunk of space debris, yet there are no planets similarly malformed.

I find it easy to conceive that the faster time-rhythm of deep space would be caught by the motion of the revolving galactic arm, condensing the light back into matter form, thus causing the time differences that are the reason matter has separated into clumps of related atoms. The bright blue haze that is seen surrounding forming stars could be the result of its locality having a faster time-rhythm with its more energetic side-to-side movement. The light would then be visible as laser light instead of a reflection off of local dust, as is currently thought. A concentration of dust would block out radiation, and does not seem capable of reflecting light across light years of space. The positively charged solar wind is drawn to an area like the Pleiades, where a negative charge exists in the newly forming star's faster time-rhythm, thus collecting matter at a faster rate.

The old Edison electric lines that conducted direct current reportedly glowed in the night. Laser light is visible and I concluded that the side-to-side path it took was responsible for this behavior. The expansion of the electron around the wire associated with the faster time-rhythm theme, which I've been promoting, links the visibility factor to the side-to-side motion.

This formation of matter would be just the opposite of a black hole, where time freezes for objects crossing the threshold but time collapses rapidly as witnessed by the outside observer. Matter can be transformed into energy, and the reverse is also theoretically possible, if the proper conditions exist. The initial manufacturing of matter, in my theories, would be the result of having a long wave–passing photon trapped by an opposing motion of ether. This would alter the path of that light similarly to starlight being acted upon by gravity, moving it to the side but pushing the trailing edge to a greater degree, resulting in a local swirl of ether that affects time. This effect is demonstrated as mass.

Conditions for the collecting of matter following this initial phase would be caused by the more massive portions developing a comparative positive charge, which would accumulate local faster time-rhythms' negative-charged matter. Matter could then slow the time scale of space by dragging it in opposing directions. There may be a rather large velocity-window for this condensation of radiation back into matter to occur. The fluctuations of the light-wave length and height may also be a factor in creating the variety of elements all at the same time. This would provide an easier-to-understand explanation of the proliferation of heavier elements, rather than having the elements be by-products of stellar evolution. Having the light-wave height influence the end product would also explain how carbon can appear in so many pure forms.

Consider the idea of a black hole as a possible model for our universe. Though this seems illogical because the time frame as we expect it would seem to be over almost instantly, time is in the eye of the beholder. At the event horizon of the black hole, all that enters is accelerated to the speed of light toward the center of the hole. At the speed of light, everything is turned into energy. With nothing but energy present, it is impossible to determine length or time. From the inside, the entire event horizon would appear as a single point. Ether racing in toward the center would be condensing into matter, which in turn manufactures ether, or in other words time fluctuations.

In a singularity, everything is one point, time is the only separation, and appearance is superfluous. A definitive explanation of this statement would be that space is irrelevant; all that space amounts to is the means of keeping track of the number of pulses that one atom is capable of achieving before it can accept energy from another atom that has dispersed that energy. I'm not going to dwell on this point but keep in mind that objects that congregate together tend to try to adjust in order to keep the same time-rhythm, as much as possible. Matter has a layered-onion time-rhythm system for its constituent particles' rhythms. Particle size is also a factor. It is common knowledge that bullet sizes between twenty-two caliber and forty-five caliber have an unstable trajectory.

According to Edwin Hubble, matter is evenly distributed through the universe in all directions, without any overabundance of matter located anywhere. This caused Albert Einstein to spend years trying unsuccessfully to construct a steady-state theory.

I subscribe to the *shrinking-universe theory* that was first tentatively suggested as a joke by Sir Edmund Whittaker, then later looked at seriously by F. Hoyle and J. V. Narlikar. This alternative theory, as strange as it may seem, would answer a lot of questions. The theory is that everything is shrinking in place, which would explain the homogenous distribution of matter throughout the universe. I include the idea that matter gives off its energy as it shrinks, which I would imagine is included in their theory. Light bending through a gravitational field, as noted earlier, moves away from a radiating body, such as the Sun, and is bent with the new angle pivoting toward the center of that mass. The red-shift would then be caused by matter, which, having been of larger size in the past, would emit a larger photon wave. The leading theory states that this red-shift is caused by the light-wave being stretched from the celestial objects moving apart.[1]

At Bell Laboratory in 1965, scientists discovered the background radiation which is the same in all directions at four degrees Kelvin. This could describe a shrinking universe, with matter making smaller photons as portions age. In this scenario, quasars would be new matter entering the universe, explaining the brightness of these stars at such a high red-shift. The quasars could then be much closer than their red-shift would seem to indicate. This is also a special case, separate from the red-shift of light emitted from an object traveling away from the observer, or light from a gravitational red-shift. The light is red due to the new atom's larger size, not as yet having been compressed to conform to the size familiar to aged matter. Preshrinking-universe theories do not leave any readily acceptable possibilities to explain these observations. Using the spectral red-shift index, these quasars would be as far away as fourteen-billion light years, and still be visible from Earth. In order to be visible at that distance, a quasar would have to be as powerful as the energy released from several galaxies exploding at the same time.

Galaxy M87 is a galaxy that has huge amounts of matter being ejected in a jet from one side. New stars seem to be forming within a bridge connecting two galaxies ESO B138 and IG 29,30.[2] Cygnus A appears to be evidence that the ejection of two large masses in opposite directions is happening from an exploding galaxy. Dr. Halton Arp at the Hale observatories photographed a bridge of gas connecting NGC 4319 and the quasar Markarian 205. The red-shift of NGC 4319 indicates a recession velocity of 1,050 miles per second; Markarian 205's red-shift corresponds to 13,000 miles per second. This is proof that the Hubble law regarding measuring distance by red-shift does not always apply.[3]

It's easier to imagine an infinite universe than a closed one because of the need for something to be on the other side of any imaginary barrier that one can conceive of. Anyone who's read this far should be familiar with Alexandre Friedmann's mathematical description of the universe. It states that for any direction you care to point, you can draw a straight line from your present location and eventually that line will bring you back, from the opposite direction, to your starting point. This caused people to calculate the quantities of matter in the universe in order to establish its size. The resultant problems were Euclidean in nature: For any line, a parallel line exists or any number of parallel lines. They all must return to their respective points of origin and still remain parallel throughout their journey. The Euclidean description of a straight line is equivalent to a ray of light, which has been observed to bend in the gravitational field of the Sun. This helps somewhat to imagine warped space. It would seem that the amount of warp also depends on the speed of the object considered.

The appearance of the universe would have to be finite, because the sky is far from being completely white in every direction. The first widely noted objection to an infinite universe was attributed to Heinrich Olbers, who wrote about this theory in the year 1823. The difficulty is that in an infinite universe, nearly every line of sight would fall on a star, which would make the night sky as bright as daylight. Some current theories limit the universe to fourteen billion light-years in radius. With my line of reasoning, two objects traveling away from each other would appear as black—light from a retreating object at light speed could not project through intervening ether, and would thus disappear from our point of observation. In this way we could eliminate impossible-to-imagine boundaries and regain an infinite universe. Light that appears black right after a nuclear implosion proves that matter can reach infinity and send out long rays while surrounded by another time-rhythm. Matter loses two of its four dimensional planes and expands spherically as a vibration wave.

An example of incompatible time-rhythms is found in a thermos. A thermos keeps liquids hot by using a double casing, with a vacuum between casings. The vacuum is a faster time-rhythm that finds the long-wave energy from the inner container incompatible, and reflects it back inward.

A demonstration of incompatible energy levels would be found in a craze that spread from California in the 80s, called *fire walking*. The size and composition of the fire pit cause the heat from the coals to resonate at a frequency that combines with the long gravity waves from below. This causes a distortion into an emulation of the waves from a faster time-rhythm, as in the belljar illustration. These waves appear to be from an object moving away.

The time spent preparing the pit allows for uniform wave length. I saw a program in which a man tried to prove that this feat was the result of the Lyden effect. The man claims that water droplets create a cushion of steam, much the same as licking your finger and checking to see if an iron is hot. He did burn himself, but not as badly as I expected. I conclude that the water that clings to his feet is not completely resonating with his aura, allowing the water to boil. This in turn burned his feet in a two-step process. A neon light also utilizes this process. The powder coating on the inside of the glass neon tube converts ultraviolet light to visible white light. For every two short photons the neon's electrons absorb, they return one longer photon.

If you were a being capable of withstanding the heat in the pit while looking at the fire walker, he would probably appear as a silvery person almost totally reflecting light. I doubt that the fire walker would appear transparent because of feeling lighter. The sensation of fire walking is reported to be that of floating on air. Without delving into this too deeply, I would imagine this would be on the order of subtractive primary colors compared to the additive ones.

In an effort to make meatloaf tastier, imagine trying to put a layer of cheese in the middle. The result of this will be that the cooking time for the meat will be much longer (the actual amount of extra time is still unknown). With the introduction of the cheese, the cheese absorbs the heat while the meat becomes semi-transparent to long-wave energy.

In the faster time-rhythm near a newly forming star, conditions that we presume to be normal might not be the same. What if you discard a rope and it is caught in orbit? It might not align perpendicular to the surface, but could be horizontal. Imagine sighting a comet approaching a star in the Plieades. I wonder if it would resemble a horizontal pancake, flattening ever larger on approach, or if it might stretch lengthwise, aligning itself to the greatest magnetic activity along the star's equator? Would the more massive portion lead toward the east or act contrary to logic and trail westward? What if this is how rings are formed around planets in the outer reaches of our own solar system?

I believe that what we perceive as color or transparency relies on the vibration qualities of the observed matter. The atom can vibrate either from side to side or in an expanding-contracting mode. As the telephone wire observation indicated, side to side would represent transparent.

The two-photon puzzle solution by John Clauser seems to be proof in favor of quantum mechanics; but, I write it off as matter that is no longer dependent on our concept of time relationships. The idea that the photon sphere can control an expanding area that grows at light speed (the photon polarity can flip simultaneously at all points) indicates the control has a steady acceleration. Suppose the photon source is vibrating vertically to your point of reference; when flipped, all points of the expanding sphere now perceive this as a horizontal vibration. I wonder what happens if one side of an expanding photon sphere experiences a polarity change at the same time the other side is experiencing a contradictory one?

I read a short time before I wrote this that scientists were going to retry the "beryllium dust in space" experiment in April of 1990. They were hoping the solar wind would excite the dust into an imitation of a comet's tail. I expected another failure. The beryllium dust is from a slower time zone.

Notes

1. William J. Nanfmann II, *Relativity and Cosmology* (New York: Harper & Row, 1977).
2. Carl Sagan, *Cosmos* (New York: Random House, 1980).
3. Nanfmann, *Relativity and Cosmology.*

Eleven
Conclusion

In conclusion, I believe the obsolete ether theory had some merit. Ether is detectable, on a large scale; it can be observed as a deviation of velocity in the orbit of Mercury. The nature of this ether is simply a measure of time differences. Ether eventually alters matter to its time/length scale, but there is a lag in the transposition. Matter is simply a pocket of ether that has partitioned itself off, and controls an area by structuring itself as an atom. The nucleus of the atom determines the atom's density through its reluctance to change, and the electron cloud is the boundary that restricts the atom from external disturbance. The atomic clock within an atom emulates a tank circuit from electronics. A tank circuit is composed of a capacitor in parallel with a coil, which is equivalent to the nucleus and electron cloud. An input signal causes the circuit to oscillate at the circuit's resonant frequency, and will maintain oscillation at this frequency for some time after the input signal is removed. This frequency is calculated by the inverse of two pi times the square root of the product of inductance times capacitance.

Heat is the presence of multiple time scales that are in the process of adjusting. Energy acquisition by an atom, and its subsequent increase of mass as a consequence, following the logic of the theory of relativity, seems contradictory. The crux of this dilemma lies in the contradiction between the concept of a slower time-rhythm being associated with a more massive object. When heated disproportionately, the heated portion of a gas or liquid will rise within a volume of identical atoms. When an atom absorbs energy, the amount of distance of a surface-oriented vibration, within the electron region, becomes greater per pulse of the atom's internal clock, which means it has acquired a faster rhythm. This energy acquisition registers as an expansion of the atom. Matter of a faster time-rhythm seeks a position farther from the center of mass. The vibration qualities of an atom are two dimensional. The expansion-contraction to side-to-side ratio of an atom's vibration determines the status matter's time-rhythm.

Imagine air passing over your arm, which causes windburn. The air is absorbing energy from your arm in a manner that compresses the escaping energy's wave height, which creates a cooling sensation while burning your arm. The moving air also causes a drop in air pressure on the surface, even though your arm is making contact with more molecules per time-pulse.

I keep separate the energy classifications of light and electromagnetic waves. These phenomena coexist as yin and yang do; one is a reaction to the other. As we observe a photon, it is merely a vibration of an atom whose message is conveyed at the speed of light through the ether. The ether is the response of the local time-rhythm to this message. The vibration, occurring as a jump from one spot to another, would project as a square wave through the ether. Each atom encountered would initially respond at its outer fringe to this intrusion by a distortion occurring in its electron's shell. The electron shell, possessing a time element, would lag in its response and appear as a sine wave to this intrusion. The long waves, being a product of the expansion-contraction cycle, would be slower because they occur in the direction of motion. The long waves are also associated with the positive charge, as when magma flows from a volcano. The lava, from the slower time-rhythm at the center of Earth, is generally accompanied by an electrical storm. The more frequent short-wave energy is representative of a faster time-rhythm. At Earth's surface, we exist in a faster time-rhythm, which is representative of a negative charge and is magnetically attracted to the positive core.

What if the slowing of the Earth's rotation is an ominous repercussion of tampering with the Earth's time-rhythm, from continued nuclear testing? The lack of protection by the ozone layer might be caused by a different rate of vibration, rather than being caused solely by chemical pollution. If this is the case, we will not cause this problem to go away by just cleaning up our chemical pollution policies.